UNITED STATES NAVAL INSTITUTE
SERIES IN OCEANOGRAPHY

OPTICAL PROPERTIES OF THE SEA
  Jerome Williams

BATHYMETRIC NAVIGATION AND CHARTING
  Philip M. Cohen

OCEANOGRAPHIC INSTRUMENTATION
  Jerome Williams

By the same author

Oceanography: An Introduction to the Marine Sciences

Sea and Air: The Marine Environment, 2nd Edition,
    with J. J. Higginson and J. D. Rohrbough

Optical Properties of the Sea

Oceanography: A First Book

Oceanographic Instrumentation

JEROME WILLIAMS

NAVAL INSTITUTE PRESS

ANNAPOLIS, MARYLAND

Copyright © 1973
by the United States Naval Institute
Annapolis, Maryland

All rights reserved. No part of this book
may be reproduced without written permission
from the publisher.

Library of Congress Catalogue Card Number: 72-92657

ISBN: 0-87021-503-5

Printed in the United States of America

To
> **D W P**

*whose initial inspiration has motivated me
throughout my professional career*

Contents

	Preface	xi
Chapter 1:	Introduction	
	What is an Oceanographic Instrument?	3
	Instrument Design Criteria	3
	Instrument Development	5
	Instrument Systems	5
	Instrument Characteristics	6
Chapter 2:	Accuracy	
	Environmental Complications	9
	Statistical Parameters	9
	Error	13
	Time Constant	14
	Other Instrument Parameters	22
Chapter 3:	Depth Determination	
	Methods of Depth Determination	25
	Pressure	26
	Pressure Sensors	27
	Signal Conditioners	30
	Pressure-Sensor Problems	30
	State of the Art	31
Chapter 4:	Temperature Measurement	
	What is Temperature?	33
	Temperature Scales and Standards	33
	Oceanic Temperatures	35
	Adiabatic Heating	38
	Accuracies Required	39
	Thermometers	40
	Pyrometers	42
	State of the Art	43

CONTENTS

CHAPTER 5: Salinity Determination
- What is Salinity? ... 47
- Salinity-Density Relationships ... 48
- Salinity Distribution ... 49
- Laboratory Determinations ... 49
- Conductivity Salinometers ... 50
- Sound Speed in Measuring Salinity ... 54
- Salinity at a Distance ... 54
- State of the Art ... 57

CHAPTER 6: Measurement of Fluid Motion
- Motion-Producing Forces ... 61
- The Equation of Motion ... 61
- Simplified Motion Equations ... 62
- Tidal Currents ... 63
- Current Magnitudes ... 63
- Vertical Motion ... 64
- Gyre, or Eddy, Size ... 64
- Measuring Practices ... 66
- Indirect Methods ... 66
- Lagrangian Direct Methods ... 68
- Eulerian Direct Methods ... 69
- Signal Conditioners ... 72
- Static Current Meters ... 73
- Turbulence Measurement ... 76
- Satellite Detection of Currents ... 77
- Current Units ... 77
- Direction Determination ... 78
- Vector Averaging ... 78
- State of the Art ... 78

CHAPTER 7: Light-Associated Measurements
- Light Losses ... 83
- Absorption and Scattering ... 83
- Irradiance and Beam Transmittance ... 86
- The Cosine Collector ... 87
- The Gershun Tube ... 87
- Wavelength Specification ... 88
- The Immersion Effect ... 91
- Relative Irradiance ... 92
- Beam Transmittance ... 93
- Irradiance-Measuring Devices ... 94
- Beam-Transmittance Meters ... 95
- The Secchi Disc ... 98
- Scattering Measurement ... 100
- Measuring Bioluminescence ... 101
- Integrating Devices ... 101
- Light as a Tool ... 101
- State of the Art ... 102

CHAPTER 8: Sound Measurements
- Sound Energy ... 105
- Absorption ... 105
- Scattering ... 106
- Spreading ... 107

	Refraction	108
	Time-Dependent Losses	111
	Acoustic Holography	111
	Ambient Noise	112
	Geophysical Sound Sources	114
	Acoustic Transducers	114
	Transducer Calibration	116
	Propagation Loss	118
	Self Noise	118
	Sound-Speed Measurement	119
	State of the Art	120
CHAPTER 9:	Chemical Measurements	
	Use of Chemical Measurements	123
	Dissolved Oxygen Distribution	123
	Oxygen Electrodes	125
	pH Distribution	126
	Indicator Methods for pH Determination	127
	Electrode Methods for pH Determination	127
	State of the Art	128
CHAPTER 10:	Measurement of Waves and Tides	
	Classical Results	131
	Deep-Water and Shallow-Water Waves	131
	Sea and Swell	133
	Wave Spectra	133
	Wave Measurements	135
	Long Waves	136
	Internal Waves	136
	Measuring Internal Waves	136
	Pressure-Sensor Wave Meters	137
	Other Wave Meters	139
	Tide Gages	141
	Measuring Waves from Satellites	141
	State of the Art	142
CHAPTER 11:	Geophysical Measurements	
	Gravity and Magnetic Measurements	145
	Acceleration of Gravity	145
	Gravity-Measuring Methods	146
	Earth's Magnetic Field	146
	Measuring Magnetic Anomalies	148
	State of the Art	148
CHAPTER 12:	Instrument Platforms	
	Fixed and Moveable Platforms	151
	The Ideal Platform	151
	Buoys	153
	Surface Ships	155
	Submersibles	156
	Scuba	158
	Satellites	158
	Aircraft	159
	State of the Art	160

CHAPTER 13:	Data Transmission and Analysis	
	Data Transmission	165
	Cable Links	165
	Acoustic Links	166
	Radio Links	166
	Laser Links	167
	Analysis Considerations	167
	Reliability	167
	Look Out the Window	168
	State of the Art	168
APPENDIX A:	Electricity Review	171
APPENDIX B:	Chlorinity Determined from Temperature and Conductivity	179

Preface

Oceanographic data are taken at sea, and anyone who has ever gone to sea knows that the marine environment is a very hostile one. Consequently, until the decade of the 1950s, most of the instruments used at sea were relatively simple and rugged mechanical devices. Facetious but often followed criteria for an oceanographic instrument were that it could contain no more than 0.5 vacuum tubes and it could be dropped on a cement pavement from a height of thirty feet and still be usable. Sophisticated measurements were accomplished by collecting water samples at sea and making analyses either in shipboard laboratories or back on the beach. An electronic instrument that did go to sea prior to that period was likely to be one being developed by a working scientist and therefore unique.

Starting in the late 1950s the development by the space program of much more reliable electronic gear spilled over into the field of oceanographic instrumentation and made possible many measurements at sea which theretofore had been impossible.

In many respects the basic principles of oceanographic instrumentation are similar to those in other fields of instrumentation, but in other respects they are very different and these differences are discussed throughout the following pages. Since the state of the art of electronic circuits and sensors is always changing, it would be a losing game to attempt a compendium of specific devices. Therefore, this book underscores only the basic concepts of instrumentation, and does not strive for completeness in listing and describing the various sensors and measuring circuits now in use. Nevertheless, many examples of sensors and measuring circuits are presented as solutions to specific measuring problems.

Generally speaking, although individual systems components are discussed, the emphasis is on the systems approach. That is to say, the entire chain, from the sensor to the final data analysis, is examined as a whole. There is also discussion of what can be measured, how it was done in the past, how it is being done now, and, when possible, how it could be done better in the future.

The author strongly believes that if the designer of oceanographic instruments is to do his job well, he must have a thorough understanding of the oceanographic environment. One of the author's aims, therefore, it to acquaint the instrument designer with some of the more important aspects of that environment. This book is neither an oceanography nor an electronics text. It is intended to serve as a bridge between the two disciplines in that, hopefully, it will motivate electronics engineers to learn more about the marine environment and acquaint oceanographers with some of the problems faced by the instrument designer.

It goes without saying that the author owes a debt of gratitude to many people who helped in the preparation of this work. In particular, the influence of Dr. D. W. Pritchard, Mr. Robert Crosby, and Mr. Edmund Schiemer must be acknowledged. Mrs. L. K. Williams is to be commended for her patience and fortitude, since she did all the typing, sometimes under conditions somewhat less than ideal.

J.W.

Oceanographic Instrumentation

CHAPTER 1

Introduction

101. What is an Oceanographic Instrument?

A bather walks down the beach, sticks her toe in the water, and calls to her companion that it is warm. At the same time an oceanographic vessel in the middle of the North Atlantic ejects an expendable bathythermograph over the side. As the small sensor falls through the water, a plot of the water's temperature as a function of depth is recorded on the deck unit aboard ship.

Both the bather and the expendable bathythermograph are oceanographic instruments. They both express an oceanographic parameter in a manner understandable to the human mind, and that is the mission of an oceanographic instrument.

An oceanographic instrument may therefore be defined as a link between the environment and the mind. This definition implies that there is some understanding of the output of this link, some information which the mind can assimilate. The bather reporting that the water is warm and the bathythermograph recording $71.3°F$ on the plot both describe the temperature, admittedly one in more quantitative terms than the other. All instruments have this one characteristic in common: they all attempt to convert some parameter into an understandable quantity.

102. Instrument Design Criteria

There are some basic differences between instruments used in a laboratory and those used at sea. Some laboratory instruments, of course, are used in the field: the chemist's spectrophotometer and the biologist's microscope are both used to great advantage. However, the ocean is not the same kettle of fish as the average laboratory.

First, the ocean is very large and a sensor or, indeed, an entire measuring system, may be hundreds or even thousands of miles away from the final data output of the system, whereas in a laboratory no great distance separates the measuring part of an instrument from its recording component.

Second, most parameters that are measured in the natural environment do not reflect homogeneity in either space or time. There is usually a great deal of variability with respect to both of these frames of reference. The question then is what to measure. The concept of an average comes into being here, and the designer usually has to decide what sort of an average will give the most information to the individual who will be using the data.

One solution to this problem is to make use, within the framework of a well-designed experiment, of an instrument already in existence, and, instead of taking a small number of measurements carefully spaced in both space and time, take a large number of random measurements in both frames of reference. When this solution is used, the instrument designer has to fit his instrument into an experiment design.

Some environmental characteristics work to the advantage of the instrument designer, others work in the opposite direction. The intermediate density of water, which allows some materials to float and causes others to sink, certainly affects the design of any device that is to be immersed in the fluid, as do temperatures low enough to cause ice to form. Still other factors include the very high pressures that exist at depth in the ocean; they have led to an entirely new area of expertise and to the development of new materials and new design concepts.

Other problems are created by the chemical composition of sea water and by biological fouling. Any material that is immersed in the oceans for a long time is vulnerable to corrosion, and devices that are so immersed tend to become attractive setting areas for all sorts of organisms, both plant and animal. There are also many documented cases of buoy mooring lines being severed by fish bites.

The oceanographic instrument designer must also take into consideration the generally poor optical qualities of the oceans. Water is not an electromagnetic medium and, even under the best of conditions, visibility rarely extends beyond one hundred feet. Most often, the user of an oceanographic instrument cannot see the device he is operating and it must, therefore, be designed to operate untended. In most cases, also, instrument packages have to be lowered from the surface and, thus, lie at the end of a rather tenuous cable thousands of feet away from the operator.

Besides those characteristics of the environment that directly affect the sensor and the type of measurement to be made, certain environmental characteristics connected with shipboard operation very often influence the oceanographic instrument designer. One of these is that the small salt particles in the ever-humid atmosphere tend to corrode electrical contacts and connections at a somewhat faster rate than is usual on shore. Of course, power supplies aboard ship are not quite as good as those found ashore, so voltage and frequency variations in AC power are likely to be of much greater magnitude than those associated with shore-based power. This may be a critical problem for the instrument designer. Furthermore, and perhaps most important of all, the vibration and platform motion always associated with ships and buoys produce effects which occasionally make even well-designed instruments absolutely useless.

One other factor that must be considered in instrument design is the use to which the resulting data are to be put. A device to be used for routine survey operations certainly will be different from one intended for the testing of mathematical models: the accuracies required and the simplicity of operation differ by orders of magnitude. Again, an instrument whose data are to be used in day-to-day forecasting will require considerations for accuracy and use altogether different from either of the above.

103. Instrument Development

Generally speaking, the development of instruments for a particular discipline follows very closely the current level of knowledge in that field of study. At first the devices are extremely crude and qualitative in nature. The sea is dark, cold, and salty, for example. Our friend the bather represents the type of instrumentation that went with that limited knowledge.

As knowledge is accumulated, instruments improve and a first attempt is made at quantification. Devices produce numbers, albeit not extremely accurate numbers, but close enough for the state of knowledge, because rough measurements are adequate for the rough theories in existence.

As theories become more and more sophisticated and the need for better and better instruments develops, later quantitative devices appear. These finer instruments often have the effect of increasing the rate at which knowledge is accumulated and a leap-frog process is set in motion: new instruments produce new data, and the new data allow the generation of new theories, which in turn, require better instruments for substantiation, and so on. Seafloor spreading and the nature of oceanic turbulence are representative of concepts that have developed as a result of instruments becoming better and better.

104. Instrument Systems

We have already described an instrument as a device intended to bridge the gap between the environment and the mind. At this time let us attempt to define an instrument, or an instrument system, somewhat more exactly.

For the purpose of this book an instrument system will be defined as a device which either implicitly or explicitly compares a selected property of the environment with a suitable standard, in an understandable manner.

An instrument system may be thought of as four separate components tied together with one or more links. These components, some of which may be missing in any particular system, are:

1. A sensor, or transducer, which converts an environmental parameter (the measurand) into an electrical, mechanical, or chemical signal

2. A translator, which translates the signal output of the sensor into a convenient form
3. An analyzer, which enhances the information content of the translator's output
4. A readout, or data display, which converts this output into understandable terms.

Sometimes 2 and 3 are lumped together and called the *signal conditioner*.

Somewhere along the line between any two of these components there may be a link, such as a wire or radio link, which serves as an avenue for transmission of the signal information.

By a combination of these four components, the environment is effectively coupled to the mind. Let us look at a typical modern instrument and see how its different components fit into this idealized scheme. Consider an instrument system to be used aboard ship, consisting of an underwater package and a deckside unit so constructed that its final output is a graph of the spatial distribution of salinity. The underwater sensor of this unit is a toroidal conductivity sensor, which will be discussed later, and the output of the toroid is fed into an electronic circuit which also has an input from a temperature sensor. The signal related to the electrical conductivity of the sea water and the signal related to the temperature are combined by the translator to form a frequency-modulated tone, the frequency of which is related directly to the salinity of the water. This salinity-dependent signal is then transmitted through the link (in this case, a one-conductor cable) to the analyzer on deck. The analyzer converts this tone into an output suitable for conversion to a graph by the readout, a chart recorder. With an additional input to this system from the gyrocompass and propeller shaft of the vessel, latitude and longitude may be programmed into the system, and three-dimensional contouring is possible.

This type of system may be classified as an automatic instrument system. It should be noted at this point that an automatic system is not necessarily an automated system. An automated system is one that uses the concept of feedback. For example, if it were desired that the rate of sampling be increased by a factor of two when salinity changes were greater than $0.1\ ^o/_{oo}$ per 100 meters, an automated system would be used. Such a system would achieve the desired result by feeding back into the input information from the output of the system, i.e., the rapid change in salinity, so that more readings per unit time could be taken. The difference between an automatic system and an automated system is the absence or presence of feedback.

105. Instrument Characteristics

There are a number of desirable characteristics that an instrument designer attempts to incorporate in any device he conceives:

1. Accuracy—An instrument should reproduce or describe the environment within known limits of error.
2. Sensitivity—An instrument should detect small changes in the measured parameter. The parameter to be measured and the environmental characteristics will determine exactly what is meant by the word *small*.
3. Ruggedness—An instrument should be able to withstand shock and manhandling and continue to operate.
4. Durability—An instrument should last a long time under conditions of minimum maintenance.
5. Convenience—An instrument should be operable without the operator having to use three hands and four eyes.
6. Simplicity—An instrument should be easy both to use and to maintain, and should not require the services of a crew of engineers whenever it is used.
7. Cheapness—An instrument should be as inexpensive as possible, for obvious reasons.
8. Understandability—An instrument should be understandable, both in concept and in the manner in which the readout is presented to the observer. It should not be necessary to subject the output of the device to an entirely different analysis system to find out what it means. The Roberts current meter, for example, has an output of punched paper tape which has to be analyzed very laboriously by hand, and the conventional bathythermograph has an output of a little glass slide which also has to be analyzed with some difficulty by hand.

An instrument should meet as many of these requirements as possible. However, there are many cases in which some of them are mutually contradictory. For example, when extreme accuracy is required, simplicity and low cost usually go out of the window. The instrument designer must determine which characteristics are most important and make the trade-offs necessary to provide them.

Sources and Additional Reading

Caldwell, D. R., et al. "Sensors in the Deep Sea." *Physics Today*, July 1969.

Fairbridge, R. W. *The Encyclopedia of Oceanography*. Reinhold Pub. Corp., New York, 1966.

Hill, M. N., ed. *The Sea*. John Wiley & Sons, Inc., New York, 1963.

Holmes, R. W. and R. J. Linn. "A Modified Beckman DU Spectrophotometer for Seagoing Use." *U. S. Fish and Wildlife Service, Spec. Rept., Fish. No. 382*, 1961.

Lion, K. S. *Instrumentation in Scientific Research.* McGraw-Hill Book Co., New York, 1959.

Myers, J. J., ed. *Handbook of Ocean and Underwater Engineering.* McGraw-Hill Book Co., New York, 1969.

Neumann, G. and W. J. Pierson, Jr. *Principles of Physical Oceanography.* Prentice-Hall, Inc., Englewood Cliffs, 1966.

Snodgrass, J. M., "Instrumentation and communications." in *Ocean Engineering,* J. F. Brahtz, ed. John Wiley & Sons, Inc., New York, 1968.

Sverdrup, H. U., M. W. Johnson and R. H. Fleming. *The Oceans—Their Physics, Chemistry, and General Biology.* Prentice-Hall, Inc., Englewood Cliffs, 1942.

USAF. *Handbook of Geophysics.* The Macmillan Co., New York, 1960.

U. S. Naval Hydrographic Office. *Oceanographic Instrumentation.* Washington, D. C., 1960.

Wiegel, R. L. *Oceanographical Engineering.* Prentice-Hall, Inc., Englewood Cliffs, 1964.

CHAPTER 2

Accuracy

201. Environmental Complications

Probably one of the most important specifications applied to an instrument is the description of its accuracy. In specifications of accuracy, terms such as *error, sensitivity,* and *repeatability* are bandied about so indiscriminately that they become almost meaningless to the uninitiated. These words have very specific meanings, and the instrument designer uses them in very narrow contexts. In this chapter we shall attempt to define and discuss accuracy in its broadest sense.

Let us begin by examining a series of measurements made upon a parameter descriptive of the natural environment. As indicated in the previous chapter, if a sensor is placed in an oceanic environment and left there for a period of time, the readings it gives will vary over this period. Measurements taken at different locations throughout an area of the ocean will also vary, depending upon where and when they were taken. The normal set of circumstances, therefore, is for an instrument to produce a set of readings none of which is identical to another, no matter how the instrument is used.

This variation is a problem in accuracy not only for the data analyzer, but also for the instrument designer, part of whose design criterion should be the type of average desired from the instrument. For example, if one wants to measure rapid fluctuations in temperature, it is necessary to design an instrument whose time constant is markedly different from that of a thermometer designed to give the average temperature over a period of an hour.

202. Statistical Parameters

In essence, we are seeking the meaning of an instantaneous measurement. Of course, there is no single definition of an instantaneous measurement because it depends upon the type of device used, how long the device has been in the environment, and how rapidly the environment is changing. For these reasons it is desirable to discuss the environment and the instrument readings in terms of some basic statistics.

If we start with a series of measurements of the same parameter, x_1, x_2,

$x_3, \ldots, x_n$, then we may describe these n measurements in terms of statistical quantities. For example, the mean $\bar{x}$ is given by

$$\bar{x} = \frac{\sum_{i}^{n} x_i}{n} \tag{1}$$

If the sample is large, we may usually assume that the measurements taken represent random variations in either the instrument system or the environment, and we may treat the measurements as forming a gaussian distribution. Such a distribution is shown in Figure 2-1, from which it may be seen that the number of measurements less than the mean is equal to the number greater than the mean.

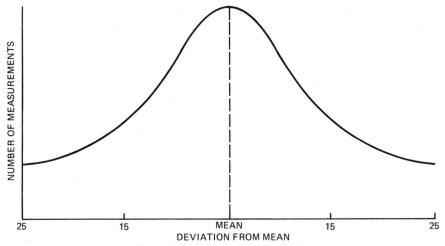

Figure 2-1. A gaussian distribution.

The spread, or scatter, of the measurements may be described in terms of the variance. The variance is given by

$$\text{Variance} = \frac{\sum_{i}^{n} (x_i - \bar{x})^2}{f} \tag{2}$$

where f represents degrees of freedom and is given by

$$f = n - 1 \tag{3}$$

Another useful statistic for measuring the spread, or scatter, of any series of measurements is standard deviation s, and its value is given by

$$s = \sqrt{\text{variance}} = \left(\frac{\sum_{i}^{n} (x_i - \bar{x})^2}{f} \right)^{1/2} = \left(\sum_{i}^{n} (x_i^2) - \frac{1}{n} (\sum x_i)^2 \right)^{1/2} \tag{4}$$

The second expression above is identical with the first, and, if a desk calculator is to be used, it is often more convenient to use the second. One standard deviation on each side of the mean represents about 68.3% of the total measurements present, so that the specification of the mean plus or minus one standard deviation will include more than two-thirds of the population.

Another method of describing the spread, or scatter, of the measurements taken involves the average of the deviations from the mean. This is called *average deviation, a,* and is given by

$$a = \frac{\sum_{i}^{n} |x_i - \bar{x}|}{n} \tag{5}$$

If it is desired to obtain a relationship between the average deviation and the magnitude of the readings being obtained, rather than the magnitude of this deviation, the average deviation may be expressed in percentage form simply by dividing it by the mean and multiplying by 100. This is called *percent uncertainty* and is given by

$$\text{Percent uncertainty} = \left(\frac{a}{\bar{x}}\right) 100 \tag{6}$$

The two most commonly used methods of error measurement are *standard error* $s_{\bar{x}}$ and *probable error* PE. Standard error addresses itself to the accuracy of the mean by measuring the uncertainty in the mean. As may be seen from the following equation, standard error is the square root of the variance per individual measurement:

$$s_{\bar{x}} = \frac{s}{\sqrt{n}} \tag{7}$$

In lieu of standard error, probable error is sometimes used, because probable error represents 50% of the measurements. Probable error is defined as

$$PE = 0.6745s \tag{8}$$

These concepts are briefly summarized in Table 2-1, wherein the half-bandwidth interval is tabulated against the percentage of measurements contained within the total population. It may be seen that an interval of one-half of a standard deviation on each side of the mean contains, for example, 38.3% of the measurements, whereas one probable error (.6745s) on each side of the mean contains 50% of the measurements.

Table 2-1 Frequency Distribution of a Gaussian Distribution

Half-band-width interval	Percentage of measurements contained
0.5s	38.3
0.6745s	50.0
1.0s	68.3
2.0s	95.5
3.0s	99.7

Occasionally, it is desired to determine whether two separate sets of measurements represent the same population. If the number of measurements is relatively small, as is usually the case in any real situation, the Student's t-distribution is a convenient way of making the determination. Essentially, the t-distribution gives a multiplying factor to be applied to the standard error which allows the determination of confidence limits. This multiplying factor appears in many tables as a function of the confidence level and the number of observations available. Table 2-2 is a small example of this.

Table 2-2 Values of t for Various Sample Sizes and Confidence Levels

Sample Size n	Confidence Level			
	90%	95%	98%	99%
2	6.31	12.71	31.82	63.66
5	2.13	2.78	3.75	4.60
10	1.83	2.26	2.82	3.25
20	1.72	2.09	2.52	2.84

By using the values of t obtained from this table we may determine the limits of the mean of a particular sample within certain confidence limits. This is done by

$$\Delta = \frac{ts}{\sqrt{n}} \qquad (9)$$

wherein Δ is the limit of the mean. A mean $\pm \Delta$ represents the range that may be expected for a given population within the confidence limits desired. For example, if a group of five measurements had a mean of 5.02 and a standard deviation of 0.03, the true mean of the total population would lie between 4.98 and 5.06 to a 95% confidence level.

Example:

$$\bar{x}_1 = 5.02$$
$$s_1 = 0.03$$
$$n = 5$$
$$\Delta = \frac{ts}{\sqrt{n}} = \frac{(2.78)(0.03)}{\sqrt{5}} = 0.04$$

From the calculation made above we know that, with a confidence level of 95%, the mean of the sample lies between 5.02 $\pm$0.04 (4.98 to 5.06). Knowing this, if a second group of measurements were made and it had a mean between 4.98 and 5.06, we could be reasonably sure (to a 95% confidence level) that the populations measured were identical. In this manner major anomalies in instrument operation may be determined.

203. Error

If *accuracy* is defined as the absence of error, then *error* is the difference between the true value and the measured value. Of course, the true value can never be known. However, the instrument designer will work within specified limits of this true value so that he will be able to describe adequately the accuracy of his instrument.

Generally speaking, errors may be divided into three groups: random errors, systematic errors, and illegitimate errors.

Random errors are statistical and represent variation obtained under essentially the same conditions. They may be caused by instrument tolerance, temperature fluctuation, small disturbances such as vibration or noise, and perhaps even the definition of the quantity being measured.

Systematic errors are errors that are inherent in the measuring system and, consequently, those from a single cause are always in the same direction. They may be caused by the instrument operator, the instrument itself, the method in which it is used, or the environment, to give a few examples.

Illegitimate errors are almost always avoidable and should not arise in quality work. They are caused by blunders in instrument reading, faulty computations, chaotic effects such as abnormally high vibrations, instrument malfunction, and so forth.

A commonly used illustration of these three types of error is that of three targets, as shown in Figure 2-2. The targets represent the efforts of three marksmen.

 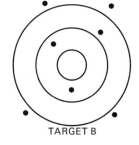 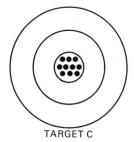

Figure 2-2. Types of error.

In the first case, A, we can see that all of the shots except one, which appears in the lower left, are bunched in the upper right-hand corner of the target. We can see also that all three types of error are represented. Random errors are represented by the size of the clump of shots; the shots did not all fall in exactly the same place but were distributed over a finite area. A systematic error is represented by the fact that most of the shots are displaced in the same direction from the center, at which it is assumed the marksman was aiming. An illegitimate error is represented by the one shot that strayed from the course and ended up far away from all the others.

The question always arises as to whether a single illegitimate error should be counted or disregarded. For practical reasons, measurements of this type are usually disregarded, but extreme care should be exercised in handling what appears to be illegitimate error, because there is always the danger of throwing away valuable information indicating important anomalous behavior.

The second target, B, illustrates an interesting situation in that the mean of these shots appears to be in the bull's-eye. The systematic error is probably quite small, but the random error is quite large. If ten shots were fired, what happened to the other three? Do they represent illegitimate errors or a large random error? Probably, in this case, they represent a large random error.

The third target, C, is, of course, the desirable one, and it is the one that very few people ever see. In this case, all the shots are grouped very closely in the center of the target: random error is small, systematic error is very small, and there does not appear to be any illegitimate error. This may be considered accurate shooting.

In addition to the classification of errors as random, systematic, or illegitimate, the terms *precision, bias,* and *accuracy* are very commonly used.

Precision, often called *repeatability,* is the agreement of repeated observations of the same quantity. It is therefore a measure of random error, i.e., the spread, or scatter. Precision is measured by standard deviation, average deviation, percentage uncertainty, standard error, and probable error.

Bias is the displacement of the observed value from the true value and is therefore a measure of systematic error.

Accuracy is a combination of precision and bias.

Target A, for example, is precise but biased; Target B is imprecise and unbiased; and Target C is accurate—that is, precise with little or no bias.

In practice, we usually take a series of measurements, rather than a single one, in the hope of avoiding illegitimate errors. With random errors canceled out, the resulting measure can be expected to be a mean, rather than an instantaneous, one ($\bar{x}$ rather than x_i).

Since systematic errors can be eliminated only by comparison with a standard, they are usually minimized by the process of instrument calibration. That is to say, a standard comparison is made for each reading of a given device so that when the device is read in the field the measurements it yields indicate little or no systematic error.

204. Time Constant

As indicated in Chapter 1, one of the problems in designing an instrument for use in the marine environment is the fact that the environment itself often changes very rapidly. A device, therefore, must be capable of following such changes closely: its response rate must be such that it can keep up with and accurately describe the changes. Thus, accuracy is not only a function of the parameters described in the previous sections, but it also related to the time constant of the device. Let us in the following pages, then, attempt to develop a mathematical model describing the fact that an instrument does not respond instantaneously to a change in the environment.

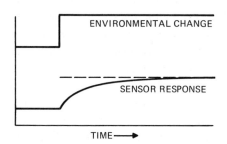

Figure 2-3. Sensor response to a step change in the environment.

If an environmental parameter changes drastically, as illustrated in Figure 2-3, the response of a device measuring that parameter turns out to be exponential, as illustrated in the lower part of Figure 2-3. This type of response is similar to that found in a simple R-C circuit wherein a battery, switch, resistor, and capacitor are placed in series, as shown in Figure 2-4. If the switch in such a circuit is closed, current immediately begins to flow as the capacitor starts to charge. The current flow through the resistor will produce a voltage across it V_R, and the buildup of the charge on the capacitor will cause a voltage to appear there also, V_C. Starting with the instant the switch is closed, we may plot the variation of voltage with time across both the resistor and the capacitor. These plots are shown in Figure 2-5, wherein it may be seen that the voltage across the capacitor as a function of time is similar to

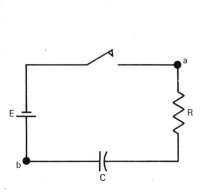

Figure 2-4. A simple R-C circuit.

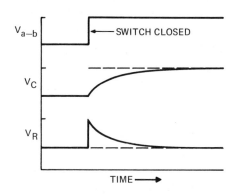

Figure 2-5. The voltages appearing across C and R in Figure 2-4.

the plot of the reading of a thermometer that is subjected to a sudden increase in temperature. Figure 2-6 shows such a plot for a thermometer that has been plunged into a hot oven. Since these two plots are so similar, an analagous R-C circuit may be used to develop equations for the time constant of any sensor.

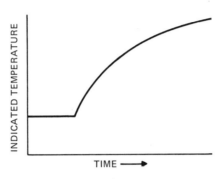

Figure 2-6. Reading of a thermometer that has been plunged into a hot oven.

An advantage of using electrical circuits in this manner is that electrical circuits may often be constructed and analogous physical parameters be determined much more simply and expeditiously than would be the case if mechanical models had to be constructed.

We shall use the circuit shown in Figure 2-4, except that in place of the battery and the switch, we will use a voltage source that is capable of varying in time as desired. Thus, our circuit will look like the one shown in Figure 2-7. Using Kirchoff's law, which states that the sum of the potential drops about any closed loop must be zero, we may write

$$E(t) = IR + V_C \tag{10}$$

and
$$E(t) = R\frac{dq}{dt} + \frac{q}{C} \tag{11}$$

where E = potential between points a and b
  I = current in the circuit
  R = resistance of the resistor
  V_C = potential across the capacitor
  q = charge on the capacitor
  C = capacitance of the capacitor
  t = time.

Since
$$I = dq/dt$$

and
$$q = CV_C$$

Then
$$\frac{dq}{dt} + \frac{q}{RC} = \frac{E(t)}{R}$$

This linear differential equation may easily be solved by the standard methods. When solved for q, we get

$$q = e^{-t/RC} \int \frac{E(t)}{R} e^{t/RC} \, dt + K e^{-t/RC} \tag{12}$$

Multiplying through by R and substituting for q we get:

$$RCV_C = e^{-t/RC} \int E(t) e^{t/RC} \, dt + KR e^{-t/RC}$$

and since $RC = \tau$, the time constant of the system,

$$\tau V_C = e^{-t/\tau} \int E(t) e^{t/\tau} \, dt + KR e^{-t/\tau}$$

or

$$V_C(t) = \frac{e^{-t/\tau}}{\tau} \int E(t) e^{t/\tau} \, dt + \frac{k e^{-t/\tau}}{\tau} \tag{13}$$

because the voltage across the capacitor is also a function of time, and the constant, KR, is set equal to a new constant, k.

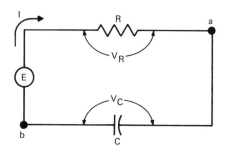

Figure 2-7. The R-C circuit used to develop the time constant equations.

Equation (13) expresses the relationship between the instantaneous value of the voltage appearing across the plates of the capacitor, $[V_C(t)]$, and the impressed circuit voltage, $[E(t)]$. In order to make use of this expression for the delineation of the characteristics of a real measuring system, let us substitute for the impressed potential difference $E(t)$, the actual value of some parameter $[A(t)]$, and for the capacitative potential let us substitute the measured value of this parameter $[M(t)]$. Thus

$$E(t) = A(t)$$

and
$$V_C(t) = M(t) \tag{14}$$

where both the actual value and the measured value of the parameter are functions of time. We then get our final expression

$$M = \frac{e^{-t/\tau}}{\tau} \int A e^{t/\tau}\, dt + \frac{k}{\tau} e^{-t/\tau} \tag{15}$$

Equation (15) expresses the measured value in terms of the time constant of the system and of the actual variation of the measurand. For design purposes, this expression is valuable because expected variations of the environment may be placed in it and it allows the measured values to be estimated. However, in the interpretation of results, one does not usually have the actual variation of the environmental parameter being measured, A. One has only the measured value, M. Thus the actual value cannot be inferred unless a mathematical expression for A may be assumed so that the integration shown in Equation (15) may be performed. However, even under these conditions, it is often possible to determine the actual value from the measured value at selected points. To illustrate this, let us go back to Equation (11):

$$E = R\frac{dq}{dt} + V$$

but since
$$q = CV \quad \text{and} \quad dq = CdV$$

$$E = RC\frac{dV}{dt} + V$$

or

$$E = \tau \frac{dV}{dt} + V \tag{16}$$

If, in Equation (16), we substitute Equation (14) to bring us into the parameters we are using, we get

$$A = \tau \frac{dM}{dt} + M \tag{17}$$

Equation (17) demonstrates a very interesting aspect of any measured record: *whenever the slope of the measured value (dM/dt) is zero, the actual value is equal to the measured value.* Thus, a record can be analyzed and selected points, at which this slope of the measured value is zero, can be found. At these points the measured value represents the true value of the environmental parameter.

Let us now analyze Equation (15) by considering three examples of an environmental parameter varying in a somewhat familiar manner and observe how these variations are reflected in the measured record.

As the first example let us assume that the actual value of the environmental parameter, A, is constant (*see* Figure 2-8). Then we have

$$M = \frac{e^{-t/\tau}}{\tau} \int A e^{t/\tau}\, dt + \frac{k}{\tau} e^{-t/\tau}$$

$$= \left(\frac{e^{-t/\tau}}{\tau}\right)(A)(\tau e^{t/\tau}) + \frac{k}{\tau} e^{-t/\tau}$$

$$= A + \frac{k}{\tau} e^{-t/\tau} \tag{18}$$

In order to determine the value of k, let us allow the measured value to be equal to the actual value at the beginning of the measuring period, that is

at $$t = 0, M = A$$

and $$A = A + \frac{k}{\tau}$$

so that $k = 0$. Therefore, for a parameter that does not vary with time, a measuring device of any time constant will give the actual value of the parameter at all times, or $M = A$ for any t.

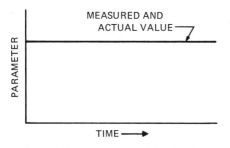

Figure 2-8. Example 1: Actual value is a constant.

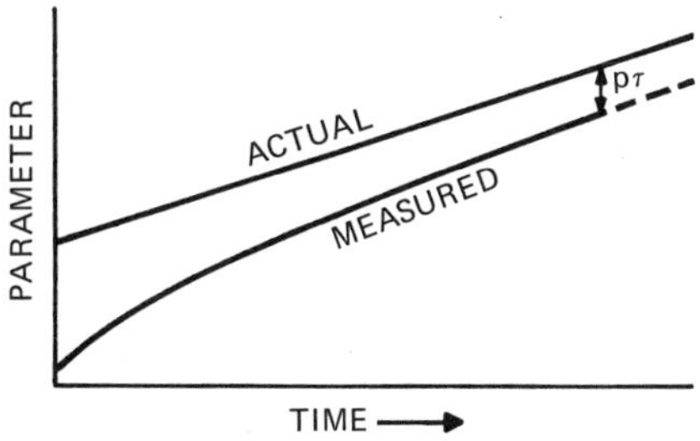

Figure 2-9. Example 2: Actual value is a linear increase.

As the second example, let us allow the actual value of the temperature to vary in a linear manner with time, as would be the case with a temperature sensor moving through a thermocline (see Figure 2-9). Thus

$$A = pt + r \tag{19}$$

Then

$$M = \frac{e^{-t/\tau}}{\tau} \int (pt + r) e^{t/\tau} dt + \frac{ke^{-t/\tau}}{\tau}$$

$$= \frac{e^{-t/\tau}}{\tau}\left((p)\left(\frac{e^{t/\tau}}{1/\tau^2}\right)\left(\frac{t}{\tau} - 1\right) + re^{t/\tau}\right) + \frac{ke^{-t/\tau}}{\tau}$$

$$= pt - p\tau + r + \frac{ke^{-t/\tau}}{\tau} \tag{20}$$

As before, to determine k, we let $M = A$ at $t = 0$

so that $$A = A - p\tau + k/\tau, \text{ since } pt + r = A$$

$$k = p\tau^2 \tag{21}$$

and
$$M = pt + r - p\tau + p\tau e^{-t/\tau}$$
$$= A - p\tau + p\tau e^{-t/\tau} \tag{22}$$

In order to express this in somewhat more usable terms, let us indicate that the measured value of temperature may always be expressed as the actual value occurring in the environment plus some incremental difference, which we shall call D. Thus

$$M = A + D \tag{23}$$

so that from Equation (22)

$$A + D = A - p\tau + p\tau e^{-t/\tau}$$

$$p = -\frac{D}{\tau(1 - e^{-t/\tau})}$$

p being the rate of temperature change with respect to time.

However, in the real environment, a temperature gradient is expressed as a change of temperature with respect to depth. If we are using an instrument that is lowered or raised through a temperature gradient, it should be a simple matter to express the rate at which this device is moving through the gradient and the gradient itself as a function of the variable, p. Thus we see that p is simply equal to the product of these two parameters

$$p = (G)(v) \tag{24}$$

where G is the temperature gradient in degrees per unit distance, and v is the speed of ascent or descent of the instrument. Thus

$$Gv = -\frac{D}{\tau(1 - e^{-t/\tau})} \tag{25}$$

and
$$v = -\frac{D}{G\tau(1 - e^{-t/\tau})}$$

If t is large, $e^{-t/\tau} \ll 1$, and

$$v = -\frac{D}{G\tau} \tag{26}$$

Equation (26) allows for the determination of the maximum rate of ascent or descent for a given instrument passing through an assumed temperature gradient under any specified error limits. It can also be used to solve for any of the other parameters that are unknown in a particular problem. For example, a descent rate may be assumed and, with a maximum temperature error, the time constant required for any particular temperature gradient may be determined.

As the third and final example, let us take an environmental parameter that is varying in a sinusoidal manner, that is

$$A = B \sin \omega t \tag{27}$$

So that

$$M = \frac{e^{-t/\tau}}{\tau} \int (B \sin \omega t) e^{t/\tau} \, dt + \frac{ke^{-t/\tau}}{\tau} \tag{28}$$

When the above expression is integrated, we get

$$M = \frac{B \sin \omega t - \tau \omega B \cos \omega t}{1 + \tau^2 \omega^2} + \frac{ke^{-t/\tau}}{\tau} \tag{29}$$

Solving for k by letting $A = M$ at $t = 0$, as before, we obtain

$$k = \frac{\tau^2 \omega B}{1 + \tau^2 \omega^2}$$

Therefore

$$M = \frac{B}{1 + \tau^2 \omega^2} (\sin \omega t - \tau \omega \cos \omega t + \tau \omega e^{-t/\tau}) \tag{30}$$

If t is considered large

$$M = \frac{B}{1 + \tau^2 \omega^2} (\sin \omega t - \tau \omega \cos \omega t) \tag{31}$$

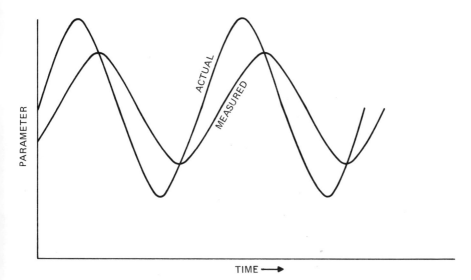

Figure 2-10. Example 3: Actual value is a sinusoid.

In order to compare the time variation of the true value with that of the measured value, Figure 2-10 shows plots of both Equation (27) (actual) and Equation (31) (measured). One interesting point revealed in this figure is that every instance of a zero slope (a maximum or minimum point) in the measured plot is a point where the graph of the measured values crosses the graph of the actual values. It is obvious that, at these crossover points, the measured value is equal and identical to the actual value of the parameter being measured, just as had been predicted from Equation (17).

205. Other Instrument Parameters

In addition to the instrument parameters of error and time constant, the instrument designer has to be concerned with four other instrument characteristics: theoretical transfer function, static error band, sensitivity, and resolution.

Theoretical transfer function is the theoretical relationship between measurand and sensor output over the entire range of operation. It is expressed as a common function, for example, linear, parabolic, logarithmic, or sinusoidal.

Static error band is a measure of how well the sensor output approximates the theoretical transfer function: it is the deviation from the theoretical transfer function. Figure 2-11 shows a linear theoretical transfer function with a static error band.

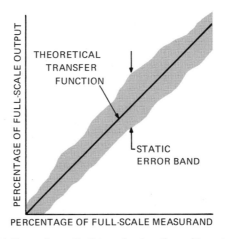

Figure 2-11. A linear theoretical transfer function with a static error band.

Sensitivity of an instrument system is the ratio of full-scale output to full-scale measurand value. Thus, a current-measuring sensor that had a range of 0–5 knots and an associated electrical output of 0–1 volts would have a sensitivity of 1 volt per 5 knots, or 0.2 volts per knot.

Resolution has to do with the smallest change of measurand that produces a recognizable change in sensor output. It is expressed as a percentage of full-scale measurand. Thus, if the current transducer above were placed in an instrument system just able to detect a change of 0.05 volts, the resolution of the system would be 0.25 knots, or 5% of full scale. The term *threshold* is often used as a synonym for *resolution*. However, it is usually applied only to the minimum value the system is capable of detecting, and therefore usually means the resolution at the low end of the scale. Generally, there is no reason to assume that the resolution will be the same throughout the instrument's range of operation.

Sources and Additional Reading

Harvey, G. J., ed. *ISA Transducer Compendium.* 2nd ed. Plenum Press, New York, 1969.

Middleton, W. E. K., and A. F. Spilhaus. *Meteorological Instruments.* 3rd ed. Univ. of Toronto Press, Toronto, 1953.

Minnar, E. J., ed. *ISA Transducer Compendium.* Plenum Press, New York, 1963.

Partridge, G. R. *Principles of Electronic Instruments.* Prentice-Hall, Inc., Englewood Cliffs, 1958.

Pingree, R. D. "In Situ Measurements of Salinity, Conductivity and Temperature." *Deep-Sea Res.,* 17(3), 1970.

Wallace, B. *An Introduction to General Instrumentation.* ISA Transportation Industry Div. Workshop Session. Plenum Press, New York, 1969.

CHAPTER 3

Depth Determination

301. Methods of Depth Determination

Of all the measurements commonly made at sea, the determination of the height of a water column is perhaps the most basic with which the oceanographer is involved. It is used both as an auxiliary in determining the position of some other parameter of interest and as a measurement by itself. We find, therefore, that depth-measuring devices are used in conjunction with all kinds of oceanographic sensors, as well as for determining changes in water depth and water level in the forms of waves and tides.

Many depth measurements are accomplished by the use of some sort of a pressure sensor. However, there are other methods.

Sonic devices are quite popular for measuring total water depth—that is, the distance from the surface to the bottom, the depth between a submerged vessel and either the upper or lower interface, and, very commonly, the distance between the bottom and devices such as underwater cameras. One problem in using sonic devices is the fact that the speed of underwater sound varies with salinity and temperature, so that, in order to obtain an accurate measurement, those parameters must be known. This subject is discussed in some detail in the chapter on sound measurements.

Although most accurate depth determinations are made utilizing pressure or sonic devices, satellites employing visible light can be used to determine some of the grosser characteristics of shallow-bottom topography. Photographs taken from satellites, especially in shallow areas where the water is clear, show the more characteristic bottom features in surprising detail. These bottom features are accentuated when special color-enhancement film and narrow-band film or narrow-band filters are used on the cameras taking the photographs. In this manner large amounts of information about shallow-ocean-bottom areas may be obtained in a relatively short period of time.

When a device is lowered at the end of a hydrographic cable, the depth of the device may be estimated by the length of cable paid out, if the wire angle is considered. Here a simple application of trigonometry produces an approximate sensor depth. Depth determined in this manner is only approximate, however, since the cable does not extend in a straight line but, as shown in Figure 3-1, curves somewhat.

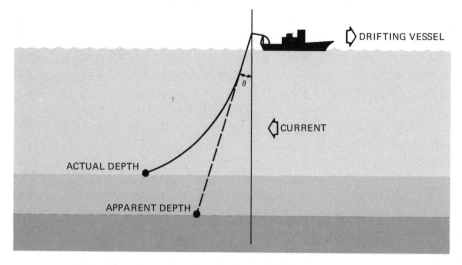

Figure 3-1. Measuring depth by using wire out and wire angle.

A method of depth determination commonly used with expendable instruments is that of predetermining the rate of the fall of the device. The expendable bathythermograph is a typical example. The unit is shaped like a little bomb whose tail fins are so designed that it spins as it falls through the water. Thus, its attitude remains constant and its rate of fall may be closely controlled. Depth is determined from the elapsed time since launch.

302. Pressure

Most instruments, however, determine depth with a pressure sensor of one form or another. This is not as easy as it sounds because the conversion of pressure to depth has many pitfalls. Referring to the hydrostatic equation [Equation (1)], it may be seen that the pressure produced by a column of fluid is related to its weight density and height,

$$p = \overline{\rho}\overline{g}h \qquad (1)$$

where p = pressure exerted by a column of fluid h units high
$\overline{\rho}$ = average mass density of the fluid
$\overline{g}$ = average acceleration of gravity within the column.

In order to determine pressure accurately, the average density of the fluid in the column and the average gravitational acceleration must be known. The experience of the bathyscaph *Trieste* provided an interesting example of how the conversion from depth to pressure can be misused. On January 23, 1960, the *Trieste,* operated by the Navy Electronics Laboratory (NEL), dived to the bottom of the Challenger Deep in the Mariana Trench in the equatorial Pacific Ocean. At the end of the dive the preliminary figure for the depth to which the vessel had submerged, as released by NEL, was 6,300 fathoms. The press release pointed out that this was an uncorrected figure, but the popular press omitted to do so and presented the figure as official. Consequently, the "fact" that the *Trieste* had made a 6,300-fathom dive was widely disseminated. Later, NEL corrected its figure to 5,970 fathoms, a difference of 330 fathoms (1,980 feet), which is certainly not negligible.

Mainly responsible for the error was the fact that the pressure gauge was calibrated in fathoms in Lake Geneva, Switzerland, where, of course, there is no salt. The difference in density between salt water and fresh water within a column of the height involved was the major corrective factor applied. The average acceleration of gravity for each layer within the column at the latitude in which the descent was made also had to be determined and applied to the uncorrected figure.

Basically, there are three sets of units commonly used to measure pressure: the engineer uses either pounds per square inch or kilograms per square centimeter, while the oceanographer uses the decibar (1/10th of a bar). The bar is defined in terms of the cgs system of units and is equivalent to a force of one million dynes acting on an area of one square centimeter. This is very close to being the pressure exerted by the standard atmosphere (1.01325 bars = 1.00000 standard atmospheres) and the pressure produced, under standard conditions, by a column of water about 10 meters high. Thus, since one decibar increase in pressure is approximately equivalent to one meter increase in depth, oceanographers find it convenient to express pressure in decibars.

303. Pressure Sensors

Deep-sea measurements of pressure are routinely accomplished with two mercury-filled glass thermometers (*see* Figure 3-2). One of these, called an *unprotected thermometer,* is similar to that used for measuring atmospheric temperatures: the other, called a *protected thermometer,* has a second glass casing with a mercury chamber to allow better heat transmission from the water to the thermometer. The two thermometers are lowered together and, being reversing thermometers, they both indicate the temperature at the time of inversion. However, the temperatures they record are found to be different. The temperature read by the unprotected thermometer is somewhat higher than that read by the protected one, because the pressure working on the unprotected thermometer tends to squeeze the mercury up into the tube a little bit more. Therefore, by standardizing the thermometers, one can determine what pressure produces a particular temperature difference between the two thermometers. When these mercury-filled thermometers are used with Nansen bottles to collect water samples, as is often the case, the reversing thermometers provide temperature and depth data concerning the sample collected.

28 DEPTH DETERMINATION

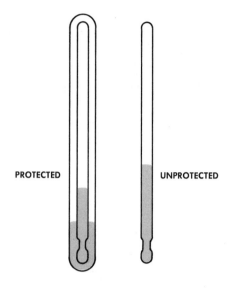

From *Sea & Air*

Figure 3-2. A protected thermometer and an unprotected thermometer.

Modern electronic instruments utilize a number of pressure sensors somewhat different from the two thermometers described above. The most common of these is the aneroid element, or the spring bellows, similar to the one used in the mechanical bathythermograph. It consists of a waterproof chamber which, under the action of water pressure, compresses against a spring in the form of a bellows (*see* Figure 3-3). As the pressure increases, the bellows is compressed and motion is experienced. This motion may be used to drive an electrical transducer or a glass slide, such as is used by mechanical bathythermographs.

Another method of producing a mechanical motion by an increase in pressure is by means of a Bourdon tube. A Bourdon tube is a curved fluid-filled tube which, under the influence of a change in pressure, changes its curvature and tends to straighten (*see* Figure 3-4). This change in curvature is the mechanical motion that can be harnessed.

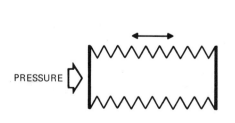
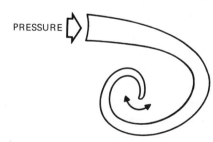

Figure 3-3. A pressure bellows.   **Figure 3-4. A Bourdon tube.**

A sensor that has been coming into more and more common use in recent years utilizes a vibrating wire as its sensing element. This vibrating wire, or Vibratron, sketched in Figure 3-5, is attached to two tines of a fork so that the frequency at which the wire vibrates is determined by the tension the fork exerts on it. When pressure is exerted on the fork, the tension on the wire changes and causes the wire to oscillate at a different frequency. The output of this device, instead of being a mechanical motion, is a change in frequency sensed by an electromagnetic pickup.

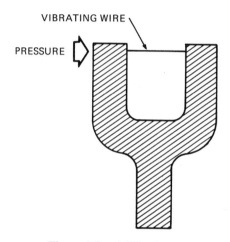

Figure 3-5. A Vibratron.

Frequency change may also be produced by changing the capacitance of an element of an electronic oscillator. This may be done by using piezoelectric elements in a capacitive configuration where an inner cylindrical piece of quartz coated with platinum film is fused to an outer section of a precision-bore quartz tube also coated on the inside with platinum film. These two elements make up a capacitor. The pressure is allowed to act on the outside tube, an increase in pressure tending to decrease the diameter of the tube, and, therefore, the spacing between the elements, so that the total capacitance of the system changes with pressure. Quartz is used because its stability is high, the elements that can be constructed with it are accurate, and temperature variations do not cause much change in its size.

There are also materials whose resistance varies with pressure. The most common of these is carbon, and carbon granules are used in some pressure sensors in just this manner. An increase in pressure decreases the electrical resistance of a package of carbon granules. Modern semi-conductors are also utilized in this mode: tunnel diodes, for example, will change their resistance with pressure.

304. Signal Conditioners

Sensor output may be conditioned in many different ways to produce the desired signal form. Some common conditioners are strain gages, linear differential transformers, coded-disc light paths, variable-frequency oscillators, potentiometers, and light-intensity photocell devices.

Strain gages may be attached either to aneroid elements or to Bourdon tubes and, since small mechanical changes cause them to vary their resistance markedly, they are very sensitive devices for measuring small pressure increments.

Linear differential transformers convert linear mechanical motions into electrical output. They may be connected directly to an aneroid element, or spring bellows, to make up a pressure transducer.

In a similar manner, mechanical motion may be utilized to turn a coded disc so that its holes line up in a predetermined manner with holes in a fixed disc. Photocell receptors are used to determine disc orientation, which may then be translated into pressure information.

As previously indicated, some sensors—for examples, Vibratrons, variable-capacitance and variable-resistance elements—operate directly into electrical circuits, driving variable-frequency oscillators. Thus, a frequency-varying output instead of an amplitude-varying one may be obtained easily. This type of output is especially suitable for use in computer inputs.

Probably the transducer most often used for converting the mechanical motion of some of the sensors described above into an electrical output is a simple potentiometer. It is commonly used with either a Bourdon or an aneroid element, in which linear motion is converted into rotational motion, and then coupled to a precision wire-wound potentiometer. The potentiometer system is a relatively simple, rugged device that provides reasonably accurate data.

A similar system uses photo-resistive elements, which are small, inexpensive, and quite reliable elements used to measure the amount of light traversing some path, as determined by the position of some mechanical flag or shield. Typically, a light shield is attached to a Bourdon or aneroid element so its position is determined by the pressure exerted on the device.

305. Pressure-Sensor Problems

None of the aforementioned transducers is perfect; they all have drawbacks. Many of them are quite sensitive to variations in temperature, the Vibratron being an excellent example of this. The rate at which the wire vibrates is related to the tension of the wire, and any change in temperature causes a change in its length and, therefore, in its tension. This drawback may be compensated for, or even offset, by placing the device in an environment whose temperature is constant, and this is usually done.

Perhaps the greatest problem in designing pressure-measuring devices is the tremendous range in pressure which many of these units have to face. Because it is often necessary to accommodate four orders of magnitude, accuracies to be expected are frequently less than desired. Accuracies on the order of 1% of full scale are not unusual, but in some cases this is not at all sufficient. This lack of absolute pressure accuracy may be overcome by using devices with limited range.

Many of these sensor-transducer combinations have extended time constants, and if they are to be employed in rapidly ascending or descending instruments, or in configurations where the pressure varies at a relatively rapid rate, they are just not suitable. Linear differential transformers, for example, often have time constants up to 30 seconds. Strain gages, on the other hand, usually have a very rapid response, the gage itself often having a time constant on the order of a few milliseconds. Thus, it is possible to design a pressure element with a short time constant.

It should be borne in mind that, for most oceanographic applications where the pressure sensor is an auxiliary device that determines the depth at which the measurements of interest are taken, the rate at which a pressure sensor is raised and lowered should determine its time constant (*see* Chapter 2). It is extremely important that instrument depths be measured as accurately as possible. The pressure sensor should not be the limiting element in the system.

306. State of the Art

The pressure-measuring devices described in the preceding sections include sensor or sensor-translator combinations which are, almost without exception, mechanical or mechanical-electronic in design. At the present time, the trend is toward the electronic end of the spectrum. New sensors just coming out of the laboratory—for example, the quartz pressure sensor (Irish and Snodgrass, 1972)—give promise of a completely electronic operational unit in the very near future.

Sources and Additional Reading

Coughran, E. H. "Some Notes on Ocean Depth Measurement." *Undersea Technol.*, 5(9), 1964.

Grasshoff, K. "Ein neuer Bodenwasserschöpfer mit Kippthermometern." *Technischer Bericht. Kiel. Meeresforsch.*, 20(1), 1964.

Hamon, B. V., D. J. Tranter, and A. C. Heron. "A Simple Integrating Depth Recorder." *Deep-Sea Res.*, 10(4), 1963.

Hyde, J. L., and E. Joscelyn. "A Temperature-Insensitive Pressure Transducer." *Mar. Sci. Instrum.*, 3, 1965.

Irish, I. D., and F. E. Snodgrass. "Quartz Crystals as Multipurpose Oceanographic Sensors—I. Pressure." *Deep-Sea Res.*, 19(2), 1972.

Kestner, A. P. "Inductional Gauge for Measuring Hydrodynamical Pressure." (In Russian). *Trudy Okeanogr. Komissii, Akad. Nauk, SSSR*, 8, 1961.

Tucker, M. J., R. Bowers, F. E. Pierce, and B. J. Barrow. "An Acoustically Telemetering Depth Gauge." *Deep-Sea Res.*, 10(4), 1963.

Welander, P., and S. Oden. "On an Instrument for Measurement of Small Horizontal Pressure Gradients." *Deep-Sea Res.*, 10(4), 1963.

Yearman, A. J. "An Ocean Depth Sensor Utilizing a Tunnel Diode as the Pressure Sensing Element." *Mar. Sci. Instrum.*, 4, 1968.

CHAPTER 4

Temperature Measurement

401. What is Temperature?

Probably the parameter in the ocean most commonly measured is temperature. It was undoubtedly the first parameter to be measured by the early oceanographers. Measuring it is normally easy, requiring the simplest type of gear, and the results are very often thought of as producing the most accurate data. However, this is not necessarily so.

No sooner does one attempt to define what is meant by temperature than one is hard-pressed for an accurate definition. Temperature is a very tenuous concept and, as we shall see, it is not accurately defined at all. The usual standard against which temperature is measured is an empirical one and has a somewhat shaky theoretical background. As a matter of fact, uncertainty in the knowledge of standard temperature is probably on the order of magnitude of $0.01°C$ and may be even greater. Nevertheless, let us attempt to define temperature as best we can. Our first attempt will be an intuitive approach similar to that of elementary science courses.

We may define the temperature of a body as "its thermal state considered with reference to its capacity for transferring heat to other bodies." Note that this concept strongly implies a relative, rather than an absolute, measurement. This implication is true of any measurement, of course, since any measurement is simply a method of comparing some parameter with some standard. However, most definitions do not imply this comparison, they define the standard.

402. Temperature Scales and Standards

An attempt at describing a standard has been made through the thermodynamic temperature scale. This is based essentially on the gas laws for an ideal gas, and therefore implies macroscopic, rather than microscopic, processes. Consequently, temperature thus defined is limited to conditions where clusters of matter contain sufficient elementary particles to provide statistical uniformity. In other words, temperature is related to the amount of material available to act on the sensor: for example, the meaning of temperatures of the upper atmosphere is somewhat doubtful.

In order to circumvent this problem, an international temperature scale has been established and adopted by most countries. This is an empirical scale, last reviewed in 1960, and it is the standard for all commonly used temperature measurements. It is based on a series of fixed points with specified methods of interpolation and extrapolation. These fixed points are supposedly easy to reproduce and therefore give an investigator a convenient means of calibrating his thermometer.

The points listed in Table 4-1 are all primary points except for the melting point of antimony which is included because it is a convenient way of interpolating between the boiling point of sulphur and the melting point of silver.

Table 4-1 Fixed Points on the International Temperature Scale (°C)

Oxygen Boiling Point	−182.97
Triple Point of Water	0.01
Steam	100.00
Sulphur Boiling Point	444.60
Antimony Melting Point*	630.50
Silver Melting Point	960.80
Gold Melting Point	1,063.00

* Not a primary point

Given a specific method for interpolating between them, these points make it possible to deduce a continuous scale. This is done by specifying both the equation to be used and the standard measuring device.

Between the limits of −182.97°C and 0.01°C, a standard platinum resistance thermometer is to be used along with Van Dusen's quartic formula for the relationship between temperature and the resistance of the thermometer.

From 0.01° to 630.50°, a standard platinum resistance thermometer is also used but Calendar's quartic relation is specified for the determination of temperature as a function of thermometer resistance.

Above 630.50°, a platinum-platinum rhodium thermocouple is substituted for the platinum resistance thermometer. Between 630.50° and 1,063.00°C, this thermocouple is used with a quartic relationship for the temperature as a function of the EMF developed, but one junction of the thermocouple must be kept at 0°C.

Above 1,063°C, temperature is determined by a narrow-band radiation pyrometer and the Planck radiation formula. Thus, a continuous temperature scale is defined running from −182.97° through temperatures higher than 1,063°.

This empirical scale obviously has its limitations. There is no defined scale below the oxygen point, for example, so that it is necessary to develop provisional scales based on reference instruments below this value. Also, there is no upper limit: a definition based on the pyrometer leaves much to be desired because proper use of the Planck radiation formula requires that the "blackness" of a black body be established, and that is difficult to do.

For oceanic measurements, temperatures between about $-2°C$ and $40°C$ are the extremes to be expected. In this range, the major limitation of the definition appears to be that the formula that relates thermometer resistance to temperature on each side of the triple point of water has to be changed. This is a minor consideration when one is working within this small range of temperature. Therefore, the international temperature scale is the one most commonly used as a standard for the measurement of temperature.

403. Oceanic Temperatures

In order to better understand the limitations of any temperature-measuring device to be used in the ocean, let us very briefly look at the temperature variations to be expected within the hydrosphere, both spatially and temporally. The temperature in the sea, including coastal and estuarine areas, varies between the approximate limits of $-2°C$ and $+40°C$. The lower limit is regulated by the freezing point of sea water. This is always greater than $-2°C$ in naturally occurring situations, and ice forms before water temperatures fall below that level. Actually, sea water having a salinity of $37°/_{oo}$ freezes at $-2.023°C$. However, water of this high salinity is never found in the high latitudes, where freezing ordinarily occurs, and in coastal areas where freezing occurs every winter salinity is usually quite a bit less.

In the open ocean the maximum temperature of surface waters rarely rises above about $30°C$. This limitation is caused mainly by the fact that over half of the solar heat energy absorbed in the upper layers of the water is used for evaporation rather than for temperature change. However, within coastal areas and areas where exchange is somewhat limited, temperatures in the summertime do rise above $30°C$ and occasionally they go as high as $40°C$.

Besides varying with season, the temperature of the upper layers of the ocean varies with latitude. As would be expected, surface water in higher latitudes are colder than those in lower latitudes. This difference does not necessarily represent a linear relationship between temperature and latitude, because the ocean surface is beset by many current systems which tend to move warm water from lower latitudes to upper latitudes and cold water from upper latitudes toward the equator. These water movements tend to distort the simple picture of temperature varying linearly with latitude. Not only are kinks in the surface isotherms produced by surface currents but also the spacing of the isotherms is somewhat distorted.

Waters in equatorial regions change temperature very slowly as latitude increases, whereas isotherms in the higher latitudes tend to be bunched together. Around the equator, for example, a temperature change of about

1°C per 10° change in latitude is fairly common, while between 40°L and 60°L the change may be 10°C per 10° change in latitude.

In addition to the bunching of isotherms, water in different parts of the world changes temperature seasonally by different amounts. As can be seen from Figure 4-1, in the region of the equator, the change in temperature is less than 1°C over a period of one year. In the region of about 40°L it is about 8°C in the northern hemisphere and 5°C or 6°C in the southern hemisphere. In the polar regions seasonal surface temperature change is relatively small: in the Antarctic Ocean it is about 1°C, and in the Arctic Ocean somewhat more.

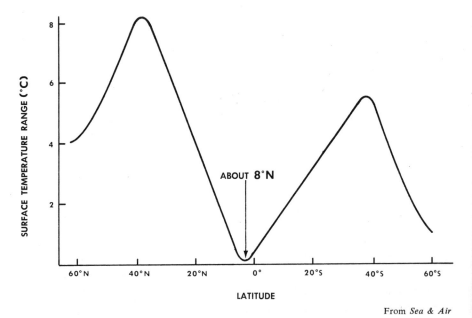

From *Sea & Air*

Figure 4-1. Seasonal temperature range of Atlantic surface waters at different latitudes.

Vertical variation of temperature in the deep ocean is also somewhat different than might be expected at first glance. Generally speaking, the ocean is composed of two layers, the upper layer being relatively warm and the lower layer relatively cold. The upper layer is sometimes called the *mixed layer,* and the lower is referred to as the *deep layer*. Sandwiched in between these two layers is the main thermocline, a region of transition between the warm and the cold water, within which the change of temperature with depth is maximal. Note from Figure 4-2 that the thermocline occurs at much shallower depths in low latitudes than it does in mid-latitudes, and that at high latitudes it does not occur at all, the cold, or deep, layer extending all the way to the surface.

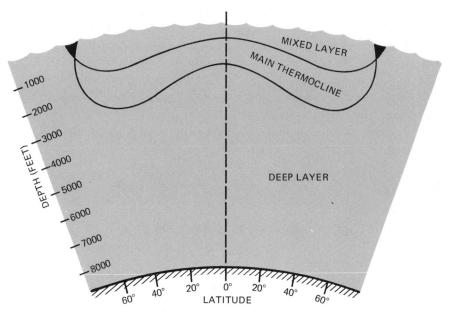

Figure 4-2. The three-layered ocean.

There is a seasonal effect in mid-latitudes that produces an additional thermocline which varies with season. This seasonal thermocline does not exist in wintertime but, as spring approaches and the upper layers of the surface waters are warmed by about 5°C, it is produced at a depth of somewhere between 100 and 150 feet. As the days get longer in the summer, more heat is added to the upper layers and the warm water tends to be mixed to somewhat greater depths. The net effect is that a warm layer, about 250 to 300 feet thick, is formed and produces a summer thermocline somewhat more marked than the spring thermocline. With the advent of fall, the surface waters are cooled but the depth of the thermocline remains about the same as it was in summer. Even though the magnitude of the temperature change involved in the thermocline is less during the fall, the vertical gradient of temperature with depth is somewhat stronger than in either of the other two seasons. As the wintertime returns and the surface layers cool off again, the thermocline once more disappears. This seasonal progression is shown in Figure 4-3.

In addition to the main thermocline that exists at depths between 1,000 and 3,000 feet and the seasonal thermocline that exists at depths between 150 and 300 feet, there is, during certain seasons of the year, a diurnal thermocline produced by daily heating. This thermocline is found at depths of about 20 or 30 feet and usually reaches its maximum about four o'clock in the afternoon. The diurnal thermocline represents a small (less than 0.3°C) temperature change, but the gradient may be large.

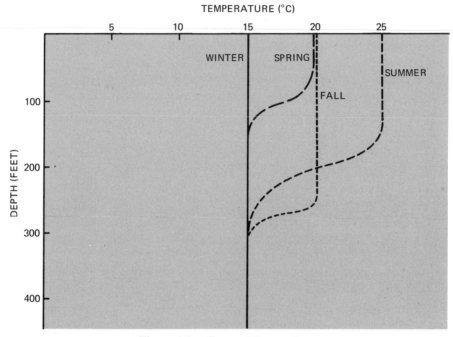

Figure 4-3. Seasonal thermoclines.

404. Adiabatic Heating

Another aspect of the vertical variation of temperature in the ocean that must be considered is adiabatic heating. Adiabatic effects in the ocean are generally small; however, when a water column as high as 30,000 feet is considered, these effects may assume great importance. Table 4-2 shows the actual and the potential variation of temperature with depth for a station in the area of the Philippine Trench in the Western Pacific. Potential temperature is the temperature a water sample would have if it were raised to the surface

Table 4-2 Deep-Water Temperature Structure

Depth (*meters*)	In situ *temperature* (°C)	Potential Temperature (°C)
2,000	2.25	2.10
3,000	1.64	1.41
4,000	1.60	1.27
5,000	1.72	1.26
6,000	1.86	1.25
7,000	2.10	1.25
8,000	2.15	1.23
9,000	2.31	1.19
10,000	2.48	1.17

without any addition or subtraction of heat. Note that the potential temperature does indeed decrease with depth. Below a depth of 4,000 meters *in situ* temperature increases as depth increases. This does not indicate a condition of instability, since the increases in temperature are caused by adiabatic heating.

Temporal variation in temperature, which is found throughout the ocean, is much more evident in the surface layers than in the deep layers, but it does exist throughout the hydrospheric environment. The hydrosphere is a dynamic system constantly in motion, and because of this water temperatures vary with time. These variations may be very rapid in the surface areas where temperature fluctuations on the order of magnitude of a few hundredths of a degree Celsius sometimes occur over time periods on the order of magnitude of a few milliseconds. There are also temperature variations whose frequencies are somewhat less. Nevertheless, it should be remembered that any temperature measurement made in the ocean is an average measurement, and that it is important to be aware of the time period over which that average is taken, as determined by the device employed.

405. Accuracies Required

The ocean then presents a somewhat anomalous situation because it is at the same time extremely turbulent and yet not homogeneous. These basic characteristics of the ocean are reflected in all parameters which are to be measured. Of course, as we leave the surface region, where most of the action is, and examine the depths of the ocean we find that the inhomogeneities are somewhat less and also the turbulence is somewhat less. This makes for readings which are somewhat more reproducible over long periods of time. Changes that occur at greater depths occur more slowly and, consequently, are of a smaller magnitude than those that take place near the surface. Much greater accuracy is required for measurements in areas where changes occur slowly and are small than for those in areas where changes occur rapidly and are great.

In order to use temperature data to determine mass movements in deep oceanic layers, it is necessary to measure temperature to an accuracy of $\pm 0.02°C$. However, in areas such as estuaries, where temperature changes are an order of magnitude greater than those in the deep ocean, temperature measurements accurate only to $\pm 0.2°C$ often serve the purpose. Furthermore, a measurement made at a selected spot in the deep ocean is probably a reasonably good spatial average of a large water mass, whereas a measurement made in an estuary, for example, is probably a good spatial average only for a relatively small water mass. The latter case is similar to the recording of meteorological data, where accuracies on the order of $1°F$ or $0.5°C$ are considered adequate for daily temperature reporting.

406. Thermometers

There are two types of devices used for measuring temperature. A device attached to or inserted in the material to be measured is called a *thermometer,* while a device located some distance away from the material to be measured is called a *pyrometer.* Both types of device are employed in the measurement of oceanic temperatures.

Most common of the thermometers used in the day-to-day measurement of oceanic temperatures is the liquid-in-glass device, consisting of a glass capillary tube filled with some liquid whose dimensions change as temperature changes. The liquid used may be an organic material such as alcohol, mercury thalium, or mercury itself. Mercury-filled thermometers, which have been commonly used in oceanography for many years, have the distinct advantage of being extremely accurate. Deep-sea thermometers usually have accuracies on the order of magnitude of $\pm 0.02°F$, or $\pm 0.01°C$. These are most often used in pairs, as adjuncts to sampling bottles, as described in Chapter 3. This system has been utilized for over 75 years in the accumulation of deep-sea data and it is still in use today.

Another commonly used type of thermometer consists of a bulb or capillary tube filled with a material whose dimensions change as temperature changes. The end of the bulb or tube is connected to a Bourdon tube in such a way that the expansion and contraction of the material within the bulb or tube produce a mechanical motion. Depending upon the expected range of temperature to be measured, liquid, vapor, gas, or mercury is generally used as the filling material.

Since the late 1930s the bathythermograph (BT) has been very commonly used to measure temperature as a function of depth for the surface layers (0–900 feet). A long capillary tube filled with xylene is used in the BT resulting in an accuracy of about 1% full scale. As may be seen in Figure 4-4, the liquid-filled tube, which is wrapped around the tail fins, is attached to a stylus that moves as the temperature changes. A pressure bellows drives a carriage upon which is placed a gold-plated microscope slide. The direction of motion of this carriage is perpendicular to the direction of motion of the

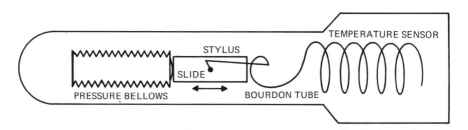

Figure 4-4. Schematic representation of a mechanical bathythermograph.

stylus so that the slide is moved in one direction with a change in depth, while the stylus moves in a perpendicular direction with a change in temperature. In this manner, when the BT is dropped, a plot of temperature *vs* depth is made to appear on the slide, which may be viewed through a calibrated viewer. This sort of a presentation is especially valuable for the immediate determination of the depth of any existing thermocline.

Although various thermocoupled devices have been used in the ocean to measure temperature, resistance thermometers are more popular. They may be of four types: the classical platinum resistance thermometer, the copper resistance thermometer, the nickel resistance thermometer, and the semiconductor thermistors, which have been increasing in popularity in recent years.

Classical resistance thermometers, be they platinum, copper, or nickel, have outputs on the order of magnitude of about 4 ohms per degree Celsius, and their outputs are quite linear in terms of the variation of resistance with temperature. Thermistors have a very much larger output, which may be upwards of 5% of its total resistance change per degree Celsius. This markedly higher sensitivity of the thermistor is mainly responsible for its popularity. However, the linear output and the basic reliability of the classical resistance thermometer make it an ideal unit for oceanographic use. Accuracies on the order of magnitude of 0.02% are possible with this type of resistance thermometer.

The thermistor, on the other hand, is not quite so reproducible, is rather non-linear in its response, and its long-term stability is not as good as that of the resistance thermometer. Experiments indicate that individual thermometers behave differently under the effects of pressure, apparently because of microscopic bubbles in the glass surrounding the thermistor material. Therefore, thermistors to be used at great depths should be individually calibrated at these high pressures.

One modern temperature-measuring device that uses a thermistor is the expendable bathythermograph (XBT). An expendable bathythermograph is a bomb-shaped device, as illustrated in Figure 4-5, containing a thermistor at its nose and a spool of thin copper wire attached to a deck unit, which records variations in temperature as a function of depth. Temperature variation is determined from the output of the thermistor, whereas depth determination is simply a function of time. Since the device is designed to fall at a constant rate, knowledge of the time of launch determines how deep it has fallen. The expendable BT has tended to replace the mechanical BT in use at sea. It allows temperature measurement at greater depths and speeds than the mechanical BT and it is also somewhat more accurate. However, the design of the system is such that within the upper 30 feet of the water surface its output is somewhat unreliable.

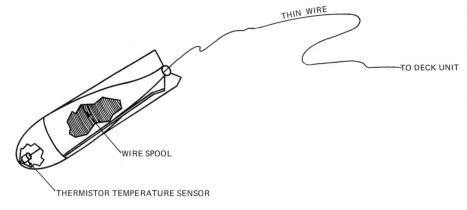

Figure 4-5. An expendable bathythermograph.

Another type of thermometer that has great promise is the quartz crystal thermometer whose resonant frequency varies with temperature. Major advantages of the quartz sensing unit are that its long-term stability is quite high, its short-term accuracy is extremely high, and its output is a varying frequency which is readily usable as a computer input. Sensitivity of a typical unit is apparently on the order of magnitude of 1,000 hertz change per degree Celsius. Its theoretical transfer function is also quite linear. Short-term accuracies on the order of magnitude of $\pm 0.0001\,°C$ have been reported, although long-term accuracies are probably on the order of magnitude of $\pm 0.01\,°C$.

407. Pyrometers

One disadvantage of using thermometers for oceanic temperature measurement is that each measurement so taken represents only a small portion of the ocean. Pyrometers make it possible to obtain data which is more representative of a synoptic picture of a large area. This is commonly done by the use of infrared thermometers which may be used either from aircraft or from satellites. These devices have an accuracy on the order of magnitude of only $\pm 5\,°F$; however, their ability to detect differences in temperature is markedly better than their absolute accuracy. Large areas may be covered by aircraft flights or satellites in a short period of time so that a synoptic picture of ocean-surface temperature may be obtained. This is especially valuable in such areas as the Gulf Stream where marked temperature differences not only exist but also determine the position of the stream.

The basic limitation of this type of device is that the infrared sensor responds to differences in radiation reaching the sensor. These differences may be due either to the temperature of the radiating body or to the emissivity of the radiating body. Radiated energy is related to the temperature by the Stefan-Boltzmann law

$$E = \sigma T^4 \qquad (1)$$

where σ is the Stefan-Boltzmann constant.

Infrared measurements would therefore appear to be linked solely to the temperature of the radiating body. However, Equation (1) was derived for black bodies only, that is, for bodies that emit according to Planck's radiation law. If a body is a gray body and emits less energy than predicted by Planck's law, a further correction has to be applied to the Stefan-Boltzmann law. This correction is called the emissivity. In other words, for a gray body

$$E = e\sigma T^4 \qquad (2)$$

where e is the emissivity.

Thus it may be seen that two bodies may emit the same amount of energy even when their temperatures are different, as long as their emissivities are not the same. As discussed in the next chapter, this fact is applied in the employment of infrared thermometers for determining surface salinities, but it must be continually borne in mind that varying emissivities tend to degrade infrared temperature data.

408. State of the Art

At the present time, many different devices are being used to take temperature measurements in the deep ocean. For measurements at depths greater than 2,000 or 3,000 feet, the most commonly used device is still the reversing thermometer. However, for shallower depths, electronic thermometers that use resistance or quartz thermometer sensors are coming into greater use. As a matter of fact, devices of this type to be used to the greatest depths of the ocean are being designed and developed and should become operational within the next few years.

Surface measurements using aircraft- and satellite-carried infrared thermometers are also commonly employed for one purpose or another. It appears that further development of infrared techniques might make it possible for a great deal of accurate data to be obtained in a very short period of time, but at present the data produced are not accurate enough to be used in large-scale forecasting.

Sources and Additional Reading

Akamatsu, H., T. Sawara, and M. Iwamiya. "Improved 1000 m Bathythermograph." *Oceanogr. Mag.,* 20(1), 1968.

Baker, H. D., et al. *Temperature Measurements in Engineering.* John Wiley & Sons, Inc., New York, 1961.

Brainard, E. C., III. The "Navi Therm"—a new device measuring sub-surface temperature underway (Summ.). *Symposium, Diffusion in Oceans and Fresh Waters.* Lamont Geol. Obs., 31 Aug-2 Sept. 1964.

Brown, D. M., J. D. Isaacs, and M. H. Sessions. "Continuous Temperature-Depth Profiling Deep-Moored Buoy System." *Deep-Sea Res.,* 18(8), 1971.

Brown, N. L., R. J. Diehl, H. B. Martin, and P. C. Stahl. "An Expendable Bathythermograph." *Mar. Sci. Instrum.,* 3, 1965.

Clark, J. R., and J. L. Frank. "Infrared Measurement of Sea Surface Temperatures." *Undersea Tech.,* 4(10), 1963.

Denner, W. W., V. T. Neal, and S. J. Neshyba. "A Modification of the Expendable Bathythermograph for Thermal Microstructure Studies." *Deep-Sea Res.,* 18(3), 1971.

Francis, S. A., and G. C. Campbell. "A Low-Cost Expendable Bathythermograph." *Mar. Sci. Instrum.,* 3, 1965.

Frassetto, R. "A Miniaturized System for Long-Term Recording of Temperature Microstructure from Anchored Buoys Having an Accuracy of Measurement of the Order of $0.01°C$" (Summ.). *Rapp. Proc. Verb., Réunion, Comm. Int. Exp. Sci., Mer Méd.,* Monaco, 17(3), 1963.

Grafa, J. B. "A Unique Tool for Undersea Thermometry: the Thermistor Chain." *Undersea Techn.,* 8(5), 1967.

Greaves, J. R., R. Wexler, and C. J. Bowley. "The Feasibility of Sea Surface Temperature Determination Using Satellite Infrared Data." *Allied Res. Assoc., Inc., NASA Contr. Rept.,* CR-474, 1966.

Gurvich, A. S., and S. T. Egorov. "The Determination of Sea Surface Temperature by Radio Wave Thermal Radiation." (In Russian) *Fiskia Atmosferi i Okeana, Izv., Akad. Nauk. SSSR,* 2(3), 1966.

Hinkelmann, H. Nachbildung der Temperaturabhangigkeit der Leitfähigkeit von Seewasser mit Plantinthermometern. *Kiel. Meeresforsch.,* 17(1), 1961.

Khundzhua, G. G. "A Highly Sensitive Resistance Thermometer for the Continuous Recording of Temperature Pulsations." (In Russian). *Vestnik. Mosk. Univ.,* (3-Fig.-Astron.), 6, 1960.

LaFond, E. C. "Towed Sea Temperature Structure Profiles." *Mar. Sci. Instrum.,* 2, 1963.

Leiss, W. J., and R. F. Trufant. "Improved Bathythermograph and the Measurement of Ocean Temperature." *Sound,* 2(5), 1963.

McAllister, E. D. "Airborne Measurements of the Total Heat Flux from the Sea Surface, Progress Report." *Mar. Sci. Instrum.,* 4, 1968.

Nan'nti, T., H. Akamatsu, and M. Iwamiya. "Further Improvement of Deep-Sea Bathythermograph." *J. Oceanogr. Soc.* Japan, 19(2), 1963.

Nemchenko, V. I. "Measurement of Sea Surface Temperature from a Ship in Motion." (In Russian). *Trudy Morsk. Gidrofiz. Inst.,* 26, 1962 Transl. Scripta Tecnica, 26, 1964.

Ogura, J., H. Tanaka, R. Kimura, K. Taira, N. Misawa, K. Ishidawa, A. Yagihashi, Y. Hara, H. Kondo, and H. Otobe. "A Comparison Between Sea Surface Temperature Measured by an Infrared Radiation Thermometer and That by the Bucket Method." *J. Oceanogr. Soc.,* Japan, 25(5), 1969.

Oshiver, A. H., and G. A. Berberian. "Sensing Sea-Surface Temperature by Airborne IR." *Geo-Mar. Technol.,* 1(4), 1965.

Pederson, A. M. An Accurate Low-Cost Temperature Sensor. *MTS Trans. Symposium, Marine Temperature Measurements,* 1969.

Pirart, M., J. Carswell, D. Oliver, and W. H. Bell. Airborne Radiation Thermometer (FRB-2). *Fish. Res. Bd., Can., Manuser. Rep. Ser (Limnol. Oceanogr.),* No. 188, 1965.

Rasmussen, R. A. "An Expendable Bathythermograph." *J. Mar. Res.,* 21(3), 1963.

———. "An Early Expendable Bathythermograph." *Mar. Sci. Instrum.,* 3, 1965.

Rumney, G. R. "A Shallow-Water Temperature Profile Recording System." *J. Conseil,* 26(3), 1961.

Shekhvatov, B. W. "Improved Model of Electrobathythermozond." (In Russian). *Trudy Okeanol. Akad, Nauk, SSSR,* 47, 1961.

Shonting, D. H., and A. L. Kadis. "The Thermiprobe: a System for Measuring Thermal Microstructure in the Sea. *"Mar. Sci. Instrum.,* 4, 1968.

Volkov, W. G. "Towed Electrobathythermograph with One-Wire Cable Connection." (In Russian). *Trudy Inst. Okeanol., Akad. Nauk, SSSR,* 47, 1961.

Wolfe, H. C. *Temperature (Its Measurement and Control).* Sci. and Ind., II, Reinhold Pub. Corp., New York, 1955.

CHAPTER 5

Salinity Determination

501. What is Salinity?

The three basic parameters for most studies in the marine environment are temperature, pressure, and salinity. The previous two chapters have seen a discussion of the measurement of temperature and pressure and in this chapter an attempt will be made to discuss the measurement of salinity.

Although salinity can be defined in a somewhat more satisfactory manner than temperature, the definition is still not as exact as might be desired. For the last seventy years there has been a continual attempt to improve the salinity definition and some progress has undoubtedly been made, as the definition has become more operational in nature.

In 1901 an international conference resolved that "by salinity is to be understood the total weight in grams of solid matter dissolved in 1,000 grams of sea water." That definition did not last too long, since it was almost impossible to determine salinity in the manner it suggested.

In 1902 Carl Forch, Martin Knudsen, and S. P. L. Sörensen made nine direct measurements of salinity and defined it as "the total amount of solid material in grams contained in one kilogram of sea water when all the carbonate has been converted to oxide, the bromine and iodine have been replaced by chlorine, and all organic matter completely oxidized." That definition also does not allow for the rapid determination of salinity and, as a matter of fact, those nine measurements are the only ones ever to be made following this method.

Although it is extremely difficult to measure salinity directly, in 1902 Knudsen was able, using the nine direct measurements, to produce an empirical relationship between salinity and another parameter called *chlorinity*, which was

$$S°/_{oo} = 0.03 + 1.805 \; Cl°/_{oo} \qquad (1)$$

where $S°/_{oo}$ = salinity in parts per thousand

$Cl°/_{oo}$ = chlorinity in parts per thousand

This was a much more usable definition of salinity since chlorinity is relatively easy to determine by direct chemical means. For example, a titration using silver nitrate with potassium chromate as an indicator gives a measurement whose accuracy is somewhere around $\pm 0.02\,°/_{oo}$. For many years, chlorinity titration was the method used in salinity analyses by the great majority of marine scientists.

In the period following World War II, it was determined that the relationship between salinity and chlorinity, as determined by chemical titration, was not quite as reproducible as the relationship between salinity and electrical conductivity. For this reason an attempt was made to redefine salinity in terms of chlorinity in such a way that old data could be used and, at the same time, to define salinity in terms of electrical conductivity. In 1969 these two definitions were generally accepted by the oceanographic community.

The new relationship between salinity and chlorinity is

$$S°/_{oo} = 1.80665\ Cl°/_{oo} \tag{2}$$

While the relationship between salinity and electrical conductivity is given as

$$S°/_{oo} = -0.08996 + 28.2972 R_{15} + 12.80832 R_{15}^2 - 10.67869 R_{15}^3 \\ + 5.98624 R_{15}^4 - 1.32311 R_{15}^5 \tag{3}$$

where

$$R_{15} = \frac{\text{conductivity of sample at 15°C and 1 atmosphere}}{\text{conductivity of water at 15°C and 1 atmosphere with salinity of } 35°/_{oo}}$$

To all intents and purposes Equation (3) is now considered to be the definition of salinity, although the intuitive concept associated with the total dissolved solids will undoubtedly be retained.

502. Salinity-Density Relationships

The determination of salinity is extremely important in many aspects of marine technology. It is, of course, important in many biological studies, but it is also important in determining the speed of sound and calculating deep-ocean currents. As we shall see later, it is impossible to measure currents at sea with even a modicum of accuracy. Consequently currents are most often calculated from a knowledge of density distribution, and, in order to determine density distribution, salinity, temperature, and pressure must be known.

Density, of course, may be measured directly by use of a hydrometer or by weighing techniques. However, these means are not satisfactory due to accuracy and convenience limitations. Currents, especially those that occur at great oceanic depths, are small in magnitude and are produced by relatively small density differences. These density differences may be caused by salinity differences on the order of magnitude of $0.02°/_{oo}$. Thus, a deep-ocean oceanographer needs to measure salinity to an accuracy of at least $0.02\,°/_{oo}$ if he is to use that measurement to determine fluid motion.

503. Salinity Distribution

The salinity of the oceans is relatively constant. More than 99% of oceanic water has a salinity between $33°/_{oo}$ and $37°/_{oo}$, with the greatest portion falling very close to $35°/_{oo}$. Of course, there are isolated areas in the ocean where one may expect deviations, sometimes quite large, from these figures. With a few major exceptions, these deviations occur at the surface and are produced by evaporation, precipitation, runoff, ice formation, ice melting, and mixing with other water masses. In coastal regions where large amounts of runoff from rivers mingle with ocean waters, salinity is likely to be somewhat less than $35°/_{oo}$, and in tropical regions where the evaporation rate is quite high salinity is likely to be somewhat greater.

Ice formation tends to increase surface salinity and, therefore, the density, so that excess salinity is apt to be readily mixed throughout the entire water column. Ice melting, on the other hand, tends to decrease salinity and to produce water of less density than that beneath it, so that the mixing process is delayed. Consequently, in polar regions the salinity of surface waters is usually somewhat less than it is in other oceanic regions.

Since temperature usually has a greater effect on density than salinity has, it often overrides the effect of salinity on density and a decrease of salinity with depth is not unusual. This is often offset at great depths where an increase in salinity is found, but occasionally even at the very greatest depths the salinity is somewhat less than that of the water immediately above it. In general, though, the vertical change in salinity between the surface and the bottom is usually less than one part per thousand. Obviously, with total variation being generally so small, deviation may be quite important in terms of water-mass motion and mixing. Consequently, a high degree of accuracy in measuring salinity is required.

In coastal regions and estuaries, on the other hand, where the salinity varies from that of fresh water up to perhaps $30°/_{oo}$, small changes are not quite so important because the range in values commonly experienced is rather large. In such areas, where the salinity may change by $5°/_{oo}$ or $10°/_{oo}$ in the course of a tidal cycle, it is not as important to measure salinities with the same degree of accuracy. In estuarine regions it is often considered adequate to measure salinity to an accuracy of $\pm 0.1°/_{oo}$.

504. Laboratory Determinations

Because, unlike the measurement of temperature, salinity does not have to be determined *in situ*, the classical method of determining it has been to capture a water sample at the depth of interest and analyze it in a laboratory, usually by chlorinity titration. The chemical titration method has now been replaced by an electrical conductivity method, but many measurements are still made in a laboratory, either on board ship or back on land.

Determining salinity by refractometer is based on the fact that the refractive index of water is a function of salinity. As may be seen in Figure 5-1, the variation of refractive index with salinity for various temperatures is quite small. Therefore, any device that purports to use this parameter for measuring salinity must be capable of making very accurate determinations of the refractive index. Two such devices have been used: prismatic instruments and interferometers. Using a differential prismatic refractometer with a precision of 3×10^{-6}, it is possible to measure salinity to an accuracy of $\pm 0.05°/_{oo}$. Interferometers, on the other hand, which have been used by the Russians and the Italians, seem to have a working accuracy of approximately $\pm 0.01°/_{oo}$.

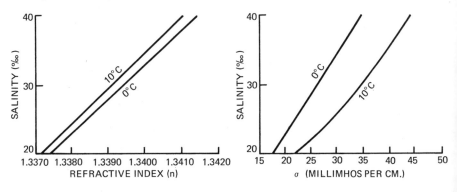

Figure 5-1. Refractive index of sea water.

Figure 5-2. Electrical conductivity of sea water.

Since these devices are usually quite sensitive to shock and vibration, they do not lend themselves too well to seagoing operations. Therefore, conductivity units have been used almost exclusively by the majority of oceanographers for measuring salinity. An additional stimulus to the use of the conductivity unit is that most manufacturers of this type of salinometer claim accuracies of $\pm 0.003°/_{oo}$, an order of magnitude greater than those claimed for the refractometer type.

Many *in situ* devices that give a continuous readout of salinity as a function of depth are now being introduced and used, however. Oceanographers are finding this type of instrument to be of great value because it enables them to study the microstructure in great detail.

505. Conductivity Salinometers

The electrical conductivity of sea water is a function of both temperature and dissolved solids. Figure 5-2 shows the variation of electrical conductivity as a function of salinity for different temperatures. Note that this variation unfortunately is non-linear, giving a different incremental value of conductivity for the same salinity at different temperatures. For this reason any device that uses conductivity as a measure of salinity must either be temperature-compensated or else must include the temperature of the sample in the salinity-determining process. The more commonly used of these procedures is temperature compensation, and it is achieved with platinum resistance thermometers, whose response curve is markedly similar to the compensation required by the conductivity relationship.

There are two kinds of sensors for measuring conductivity from which salinity may be determined: the electrode type, and the electrodeless, induction toroidal type.

The electrode sensor requires bare metal electrodes to be immersed in the sample and the electrical resistance path between them is then determined. Unfortunately, small fouling problems cause uncertainties on the order of magnitude of the accuracies desired at great depths, so this cell does not hold too much promise for use in the deep ocean. However, it is a very much simpler and somewhat cheaper type of cell than the toroidal design and, if it is properly cared for, its accuracy is good enough for use in shallow water. Consequently, many instruments for estuarine use have this type of sensor.

The toroidal sensor is electrodeless: no bare metal touches the water. For this reason, it is to be preferred over the electrode sensor when high accuracies are desired. Corrosion and fouling are minimized by the very nature of its design. In principle the cell consists of two transformers. The primary of the first transformer is the exciting toroid, a doughnut-shaped core wound with a coil and excited by an AC signal. The second transformer is the signal toroid whose secondary is similar in appearance to the exciting toroid. The output of this coil is the signal. Coupling the two coils together is a loop of water that serves as the secondary of the first transformer and the primary of the second transformer. In this manner, any electrical resistance in series with this loop causes a different amount of voltage to be coupled to the signal toroid. Of course, the varying resistance within the loop is simply the varying electrical conductivity of sea water. The equivalent circuit is shown in Figure 5-3. The two coils are usually mounted one atop the other so that

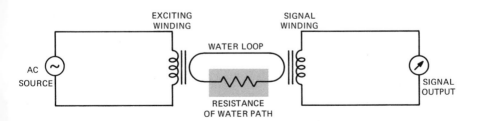

Figure 5-3. Equivalent circuit for a toroidal conductivity cell.

their common hole serves to delineate the loop of sea water. This is shown schematically in Figure 5-4. Although commercial units usually have much more elegant circuits, the basic principle outlined above is applicable to all toroidal sensors.

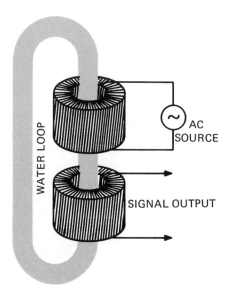

Figure 5-4. Schematic representation of a toroidal conductivity cell.

Toroidal sensors are used in both laboratory and submersible instruments. A typical laboratory unit is shown in Figure 5-5, wherein a standard seawater sample is measured and an unknown sample is then compared with it. The unit requires frequent restandardization with standard sea water.

A submersible unit, on the other hand, reads directly in salinity and does not require constant restandardization. A unit of this type, the STD (salinity-temperature-depth), is shown in Figure 5-6.

One difficulty with devices such as the STD, where salinity is measured *in situ*, is that conductivity is a function of pressure in addition to being related to the salinity and the temperature. The relationship between conductivity, pressure, and temperature is

$$\frac{\Delta G}{G} = (ap^2 + bp)(ce^{d/T} + f) \tag{4}$$

where a, b, c, d, and f are constants

G = conductivity at atmospheric pressure
p = pressure
T = temperature in °K.

Thus, STD units must be designed to allow for pressure effect.

SALINITY DETERMINATION

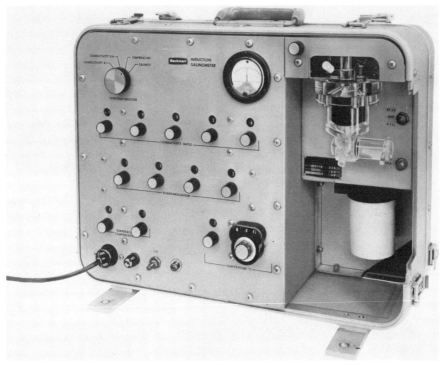

Beckman Instruments, Inc.

Figure 5-5. A laboratory salinometer.

Plessey Environmental Systems

Figure 5-6. An STD.

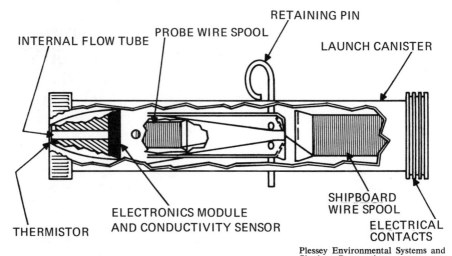

Figure 5-7. The expendable STD in cross section.

506. Sound Speed in Measuring Salinity

Another suggested method of measuring salinity *in situ* is by using sound-speed measurements. Since sound speed varies with both temperature and salinity, if temperature and sound speed are known, it should be possible to calculate salinity directly. Sound speed also varies with pressure, but that variation is very well known. Because devices for measuring sound speed and temperature are readily available, this seems like a very plausible suggestion. The only problem is accuracy.

Although the variation of sound speed with changes in temperature and salinity are related to the magnitude of the temperature and salinity when the change takes place, average values may be represented as 3 meters/sec/°C and 1.3 meters/sec/°/$_{oo}$ salinity. Thus, in order to detect a change in salinity of $0.01°/_{oo}$, it is necessary to measure a sound-speed change of 0.01 meters/sec and to maintain the temperature at a constant value, ±0.002°C. Of course, if the temperature is measured to that accuracy the same result will be obtained. At the present time it is possible to measure sound speed, at best, to only ±0.1 meters/sec and temperature to only about ±0.01°C, neither of which is good enough to obtain the salinity accuracy required by deep-sea oceanographers. For this reason, sound speed is never used to determine salinity.

507. Salinity at a Distance

If it is desired to determine the surface salinity of a large oceanic area in a short period of time, one method that is becoming available is the use of

SALINITY DETERMINATION 55

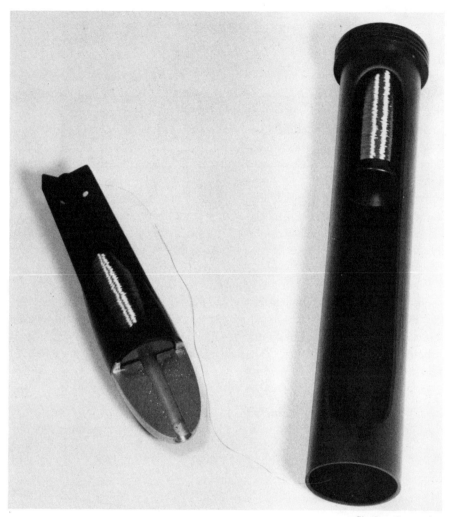

Figure 5-8. An expendable STD.   Sippican Corporation

electromagnetic waves of one sort or another. Infrared has been suggested for use by either aircraft or satellites, but work on microwave radar wavelengths has actually been done. In the previous chapter, it was shown that the energy emitted by gray body is a function not only of its temperature but also of its emissivity [*see* Equation (2), Chapter 4]. This fact is used in determining salinity from energy emitted in the microwave region of the electromagnetic spectrum. It turns out that the most sensitive frequency is in the range of 1–1.5 GHz.

In this frequency range the energy emitted by the ocean surface is picked up by the sensor and analyzed in terms of a knowledge of the actual temperature of the water surface. The temperature of the water surface is determined by a similar type of measurement at a different wavelength where the emissivity is different for the ocean surface. Figure 5–9 shows the change in apparent brightness temperature (the temperature given by the Stefan-Boltzmann law for maximum energy emission) as a function of surface salinity. It can be seen that, for salinities around $35°/_{oo}$, a salinity variation of $5°/_{oo}$ produces a difference in apparent temperature of about $2°$.

Unfortunately this phenomenon is also a function of the actual temperature of the water surface. The value of emissivity as a function of salinity for different temperatures is shown in Figure 5-10. Note that at very low temperatures, such as might be experienced in polar regions, the change in emissivity for a large change in salinity is quite small. However, for middle and low latitudes this scheme appears to hold a great deal of promise in determining surface salinities over large areas. This is especially true where large differences in salinity are to be measured—at the mouth of a large river, for example.

It should be borne in mind that, even though this type of measurement results in large amounts of data in a short period of time, the data are representative of the surface (the first centimeter or so) regime only. If information is desired about the salinity of waters below the surface, instruments have to be dropped into the water. Thus, it appears that the STD type of device will be with us for many, many years.

508. State of the Art

Although there are many ways by which salinity can be determined, by far

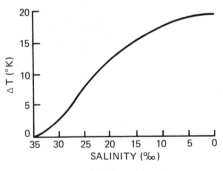
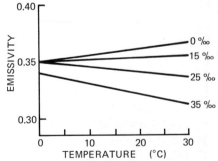

Figure 5-9. Calculated change in apparent brightness temperature for varying salinities at a frequency of 1.42 GHz. (After Droppleman et al., 1970)

Figure 5-10. Calculated emissivity change with temperature for various salinities at a frequency of 1.42 GHz. (After Droppleman et al., 1970)

the greatest number of determinations made today are done with electrical conductivity sensors. Laboratory salinometers have become relatively standardized and any foreseeable changes appear to be in the areas of reliability and durability. *In situ* units, on the other hand, are constantly being improved, even though the basic conductivity sensor remains virtually unchanged. The major improvement appears to be in extending the depth capability of the STD (salinity-temperature-depth) instrument so that salinity and temperature profiles can be obtained to the greatest ocean depths, and the expendable STD (XSTD) shown in Figures 5–7 and 5–8 seems to be a step in that direction.

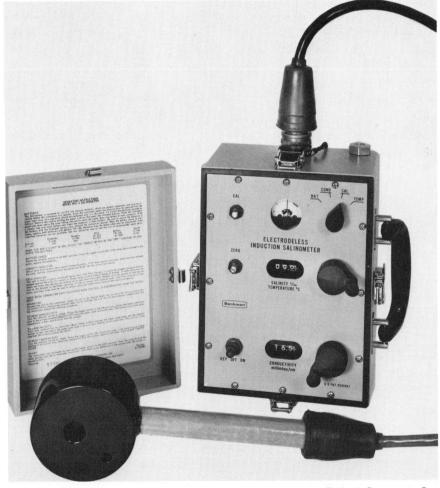

Beckman Instruments, Inc.

Figure 5-11. A portable salinometer.

Sources and Additional Reading

Antonova, L. M., and V. I. Zaburdayev. Induction salinometer with binary code readout. (In Russian) In: *Metody i Pribory dlya Issledovaniya Fizicheskikh Protesessov v Okeane,* A. N. Paramonov, ed., Izd-vo Naukova Dumka, Kiev. Translation: JPRS: 39, 13 Feb. 1967 (Clearinghouse Fed. Sci. Tech. Info., U. S. Dept. Commerce).

Brown, N. L. "A Proposed In Situ Salinity Sensing System." *Mar. Sci. Instrum.,* 2, 1963.

———. "The Paraloc®, a Precise Telemetry Subcarrier Oscillator." *Mar. Sci. Instrum.,* 4, 1968.

———. "An In Situ Salinometer for Use in the Deep Ocean." *Mar. Sci. Instrum.,* 4, 1968.

Bruevich, S. V. "Main Salt Composition and Salinity Determinations According to Knudsen." (In Russian). *Trudy Inst. Okeanol., Akad. Nauk, SSSR,* 47, 1961.

Cox, R. A., et al. "Chlorinity, Conductivity and Density of Sea Water." *Nature,* 1962.

Crease, J. "Determination of the Density of Seawater." *Nature,* Lond., 233 (5318), 1971.

Dauphinee, T. M. "In Situ Conductivity Measurements Using Low-Frequency Square Wave AC." *Mar. Sci. Instrum.,* 4, 1968.

Droppleman, J. D., R. A. Mennella, and D. E. Evans. "An Airborne Measurement of the Salinity Variations of the Mississippi River Outflow." *J. Geoph. Res.* 75(30), 1970.

Kasten, H. "Ein automatisch registrienendes Temperatur- und Salzgehaltsmessgerät für Messungen in situ 500 m." Tiefe. *Beitrage z. Meeresk.,* 9, 1963.

Lebout, H. "Dosage de la salinité de l'eau de mer par une méthode combinant la titremetrie et la gravimetrie." *Cah. CERBOM, (Centr. Etudes Recherch. Biol. Océanogr. Médic.),* Nice, 1, 1964.

Lovett, J. R. "Determination of Salinity from Simultaneous Measurements of Sound Velocity, Temperature and Pressure." *Limnol. Oceanogr.,* 13(3), 1968.

Park, K., and W. Burt. "Electrolytic Conductance of Sea Water and the Salinometer (2)." *J. Oceanogr. Soc.,* Japan, 21(3), 1965.

Reeburgh, W. S. "Measurements of the Electrical Conductivity of Sea Water." *J. Mar. Res.,* 23, 1965.

Siedler, G. "On the In Situ Measurement of Temperature and Electrical Conductivity of Sea-Water." *Deep-Sea Res.,* 10(3), 1963.

Stirling, P. H., and H. Ho. "Determining Density." *Industr. and Engng. Chem.,* 53(10), 1961.

Thayer, O. E., and R. G. Redmond. "Budget Salinity Recorder." *Limnol. Oceanogr.*, 14(4), 1969.

Volkov, W. G. "An Instrument for the Salinity and Temperature Depth Measurements. Chemistry of the Waters and Sediments of the Seas and Oceans." (In Russian; English abstract). *Trudy. Inst. Okeanol., Akad. Nauk, SSSR*, 67, 1964.

Williams, J. "A Small Portable Unit for Making In Situ Salinity and Temperature Measurements." *Trans. Instrum. Soc. Amer.*, 1962.

Wilson, T. R. S. "A Portable Flow-Cell Membrane Salinometer." *Limnol. Oceanogr.*, 16(3), 1971.

CHAPTER 6

Measurement of Fluid Motion

601. Motion-Producing Forces

One of the characteristic properties of the world ocean is the fact that it is continually in motion. Not only is the ocean as a whole continually circulating, but every portion of it seems to exhibit some sort of flow pattern. The ocean is dynamic and turbulent, but, strangely enough, it is heterogeneous, so that motion found at one point in time and space is not usually duplicated at some other point. This turbulence makes fluid motion extremely difficult to describe and thus, as will be seen, presents one of the basic problems in measuring currents.

Let us look first, though, at the forces that produce motion in the fluid environment. In general, currents are produced either by mechanical action of the wind or by thermal activity in the form of uneven heating. Uneven heating not only generates areas of unequal temperature, it also creates differing evaporation rates. Thus, salinity is changed at varying rates and that fact, in turn, produces density variations in much the same manner as those produced by changes in temperature. Since pressure is related to density, any changes in density we find in the three-dimensional ocean produce corresponding changes in pressure. It is this pressure distribution that is the driving force for many oceanic currents.

Once motion is initiated, other forces that tend to influence the magnitude and the direction of the resulting current come into play. These forces include internal friction resulting from turbulent stresses, coriolis force produced by the rotation of the earth, and centrifugal force caused by, or reflected in, currents that have a non-linear (curved) path.

602. The Equation of Motion

If the currents in the ocean are to be described in terms of the forces that produce them, it may be done by utilizing Newton's second law. The forces acting on an individual parcel of water may be summed and equated to the acceleration experienced by the parcel. This is done in

$$F_s + F_p + F_c + F_g + F_f = a \qquad (1)$$

where F_s = centrifugal force per unit mass acting on a water parcel
 F_p = pressure gradient force per unit mass acting on a water parcel
 F_c = coriolis force per unit mass acting on a water parcel
 F_g = gravitational force per unit mass acting on a water parcel
 F_f = frictional force per unit mass acting on a water parcel
 a = acceleration of the water parcel.

Note that each of these forces is expressed in terms of the force per unit mass so that the units of each are those of acceleration.

In Equation (1), the frictional force per unit mass F_f includes both the frictional force of the wind on the water surface (the tangential wind stress) and the internal friction of the water usually referred to in terms of turbulence, or Reynolds stresses.

603. Simplified Motion Equations

If each of the forces mentioned above is expressed in terms of other measurable parameters, the resulting relationship is a rather complex partial differential equation that cannot be solved in closed form. However, the gross aspects of the flow may often be adequately described if some of the terms in Equation (1) are neglected. For example, an approximation very commonly used in the description of oceanic currents is the *geostrophic approximation*. For geostrophic flow, it is assumed that all forces, except the pressure gradient force and the coriolis force, are negligible and that no accelerations are experienced. In other words, for geostrophic flow there is a simple balance of forces between the pressure gradient force and the coriolis force, as indicated in the geostrophic equation

$$F_p + F_c = 0 \qquad (2)$$

Other approximations can be made. For example, if the motion is allowed to be in a curved, rather than a straight, line, it is necessary to include the effect of centrifugal force. This motion is called *gradient flow*, which may be represented by

$$F_p + F_c + F_s = 0 \qquad (3)$$

A third approximation often discussed by oceanographers is that which results in the Ekman spiral, wherein the only important forces considered are wind stress and the coriolis force. This may be written

$$F_c + F_f = 0 \qquad (4)$$

As may be seen, all sorts of combinations are possible. However, only those which approximate conditions experienced in nature are useful. Geostrophic, gradient, and Ekman spiral conditions are often approximated in nature, so these three types of flow are quite useful.

604. Tidal Currents

Currents are produced not only by uneven heating and by changes in density caused by salinity variations, but by another source that is of extreme importance in many areas. This, of course, is the tidal force. Tidal currents differ in one important respect from all other types of currents that have been considered in that they are periodic, and predictably so. They change both magnitude and direction in a known manner because the basic force that produces them is the net gravitational attraction of the sun and the moon and its variation as the earth rotates.

However, the geography and topography of the area being considered also play a part in the adequate description of tidal currents. The Bay of Fundy, for example, has extremely large tidal currents, on the order of magnitude of 8 knots, while the European Mediterranean has almost negligible tidal currents. Not only is the Mediterranean the wrong size to resonate at tidal frequencies, but the entrance to it through the Strait of Gibraltar is too small for any appreciable volume of water to be exchanged on a diurnal or semi-diurnal basis. The Bay of Fundy, on the other hand, is so configured that it acts as a funnel for the tidal energy moving up it, and this venturi effect seems to increase tidal currents to extremely large values.

In any event, if the description of total current experienced at any particular place is to be complete, tidal currents must be taken into account. Therefore a solitary measurement, especially in a coastal area, will not be descriptive of the current to be expected because it will probably change within a few hours.

605. Current Magnitudes

Surface currents may reach values greater than 5 knots but the average is somewhat less than 1 knot. Higher values are almost always found near the shore and are typically tidal in nature, although non-tidal currents sometimes run as high as 6 knots. Higher speed currents are also generally surface currents. This is not always true, but certainly most of the large-magnitude currents found in the ocean are at, or near, the surface.

With some exceptions, the flow rate of currents existing below the oceanic surface layers tends to decrease with depth. Among the exceptions are Gulf Stream and equatorial undercurrents. As against the range of 1 to 5 knots for surface currents, deep-ocean currents are on the order of 0.02 to 0.2 knots, or 1 to 10 centimeters per second. These very small water motions are extremely difficult to measure. This is another of the problems associated with the measurement of oceanic currents, because, even though the motions are small, the total volume of water being transported is very large. Thus, a small error in speed determination results in a large absolute error in the calculation of volume transport.

606. Vertical Motion

The motions discussed above are horizontal, but there are also vertical motions within the oceanic environment, some of which are of extreme importance. For example, sinking water, which usually occurs in convergent areas where currents come together, is important in the formation of new water masses and is an essential part of the general circulation of the three-dimensional ocean. On the other hand, there are well-defined regions in the oceans where water rises to the surface. Many of the rising waters bring to the surface nutrients that aid in the growth of large plant blooms, which, in turn, provide fodder for various animals, including some that are important commercially. This process occurs off the west coasts of both North and South America, for example.

Both horizontal and vertical motions, then, are of some interest. However, except for the specific instances mentioned above, vertical motions generally are very small, usually 1 to 3 orders of magnitude less than the horizontal motions found in the same area. For this reason vertical motions are usually ignored when the current structure of a particular oceanic area is described mathematically. However, in some coastal and estuarine regions, vertical motions are, at most, 1 order of magnitude less than the horizontal motions and therefore cannot be ignored. Almost all current meters presently in use are designed to measure horizontal motion only.

607. Gyre, or Eddy, Size

The better-defined surface currents found in the ocean provide a relatively good picture of the general types of currents to be expected throughout the three-dimensional ocean. As seen from Figure 6-1, the most obvious characteristic of the surface currents is typified by such large systems as the one in the North Atlantic that is composed of the Gulf Stream, the North Atlantic drift, the Canary current, the North Equatorial current, and the Florida current. These currents form a gyre, or eddy, about 2,500 miles in diameter, and there are similar gyres throughout the world's oceans. Since these gyres are large and it takes quite a long time to circumnavigate the system, they are usually referred to as *permanent currents*. However, an examination of the ocean shows the existence of gyres only a few hundred miles in diameter and with much decreased periods; small-scale charts reveal the presence of even smaller ones. As a matter of fact, it appears that there is a continuous spectrum of eddies extending in size down to that of molecular excursions. Thus we see that the ocean contains current eddies varying in both size and period by many orders of magnitude.

Here, then, is a major difficulty in measuring ocean currents: the determination of the averaging distance and time for which the measuring instrument is designed. Obviously a device designed to measure the average daily current

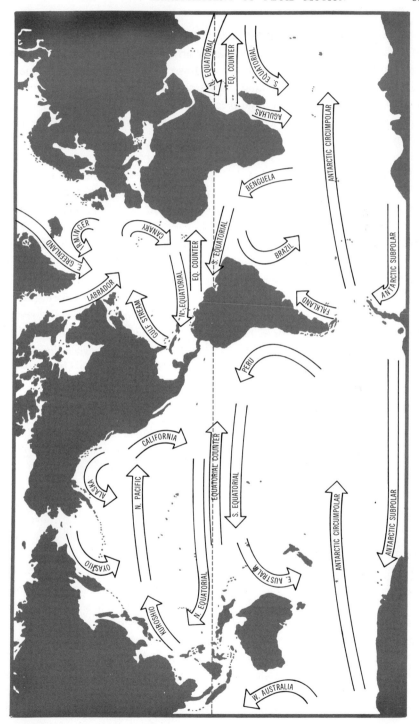

Figure 6-1. Surface currents of the world during the northern hemisphere winter.

over a square mile of the ocean surface will give a markedly different result from one designed to measure the average current over a square centimeter of the ocean surface for a period of one second. Not only must basic constants, such as time constant and threshold value, be decided, but also limitations of the instrument, including the final analysis of data, must constantly be considered.

608. Measuring Practices

Problems often arise when the magnitude of the water motion to be measured is small compared with motion present from other sources. With some sensors, the heaving motion of the ship or buoy to which the current meter is attached produces an erroneous indication of horizontal current, even though only a vertical motion of the device is induced. Furthermore, any attempt in the deep ocean to obtain a fixed reference point from which to reference motion is almost impossible. Anchoring a ship in the deep ocean so that it will not move with respect to the earth's surface, or even anchoring a buoy so that it is fixed in space, is a task which, at this time, presents insurmountable difficulties. In the deep ocean it is extremely difficult to determine the motion of the water relative to the earth's surface even when known ship drift is programmed into the calculations. This is because ship drift is not usually known accurately enough to result in a set of measurements whose standard deviation is less than the order of magnitude of the known error: the uncertainty in the readings often turns out to be greater than the magnitude of the readings themselves.

In shallow water, however, it is possible to maintain a fixed platform from which to hang a current meter, and instruments for this application have been built and used with some success.

Oceanic currents are generally determined by either an indirect or a direct method, the former being the most commonly used. An *indirect method* measures some property of the medium in order to infer flow; a *direct method* uses a device that measures the flow itself.

609. Indirect Methods

Some of the forces described in the simplified forms of the basic equation of motion [Equation (1) above] may be determined independently of any direct determination of motion, and from the results speed may be calculated. For example, if the relation for geostrophic flow [Equation (2)] were written in terms of more detailed descriptions of the forces, a typical presentation would be

$$-\alpha \frac{dp}{dx} + 2\Omega v \sin \phi = 0 \tag{5}$$

where α = reciprocal density
Ω = rotational speed of the earth
ϕ = latitude
v = speed in the y direction
$\dfrac{dp}{dx}$ = pressure gradient in the x direction.

The first term in Equation (5) is the pressure gradient force per unit mass F_p, while the second is the coriolis force per unit mass F_c. Note that, if the other parameters are known, this equation can very easily be solved for the speed in the y direction. In order to determine its value the pressure gradient must be converted into a density gradient, because pressure gradients in the ocean cannot be measured directly. When that conversion has been made, it becomes possible to calculate the geostrophic current simply by measuring the density distribution within the ocean. Probably, more oceanic currents are determined by this method than by any other.

Another indirect method that has had some success is the electromagnetic method. Here, it is not the distribution of density within the fluid that is measured, but the gradient of the electrical potential produced by a moving conductor (sea water) within the earth's magnetic field. This is done by means of a device called a *geomagnetic electrokinetograph* (GEK), developed by W. S. von Arx at Woods Hole. Very simply, the electrical potential developed by the moving sea water within the earth's magnetic field is determined by the use of two electrodes placed within the volume of the water. The potential developed V is then given by the expression

$$V = k(lH_zv) \times 10^{-8} \text{ volts}$$

where k = some constant equal to about 1.1
l = the electrode separation in centimeters
H_z = the vertical component of the earth's magnetic field in cgs units (Oersteds)
v = the magnitude of water current in cm/sec.

Thus
$$v = \frac{V \times 10^8}{klH_z} \text{ cm/sec} \qquad (6)$$

The relationship given in Equation (6) may be used for measuring stream transport or ocean currents, the only basic difference being in the value of k.

A GEK is usually used to measure surface currents, although it has been used down to depths as great as 500 meters. It appears, however, that this method is more reliable as an indicator of current direction than of speed, and for this reason it is sometimes used in conjunction with the geostrophic method to determine current direction more accurately.

There are many sources of inaccuracy in an instrument of this type, most of them involving the electrodes. For example, the electrodes are quite unstable, varying in output by about an order of magnitude over a relatively short period of time. Also, it is impossible to construct them so that they are exactly the same, and local differences in temperature, salinity, and dissolved oxygen can cause the contact potential of each to differ by as much as 1 millivolt. This requires a zero determination before each reading, and that is made by averaging the two readings obtained by trailing the electrode pair over two courses at right angles to each other. In recent years the GEK has not been used as much as it was in the past; however, the concept of determining water flow by measuring induced potential is valid, and instruments based on it keep reappearing in different forms.

610. Lagrangian Direct Methods

In contrast to an indirect method of obtaining flow by measuring some property of the medium from which the current may be calculated, is the direct method where the motion itself is measured. There are two types of motion-measuring devices. The Lagrangian, or drifting object, is the first of these. Current magnitude and direction are obtained by noting the sequential position of the device as a function of time. In contrast to this, there is the flow type of device that is fixed in space and measures motion past it. This is sometimes referred to as an Eulerian type.

Drift devices are probably the oldest instruments used by scientists for measuring currents. The classical type, of course, consists of a stoppered bottle containing a note urging the finder to communicate the time and place of his discovery: if the finder complied, the trajectory of a water parcel could be determined. This general idea is still used today, both bottles and smaller plastic cards being used. The cards are used in an attempt to get as much of the cross-sectional area of the drifting object as possible out of the influence of the wind and into the influence of the moving water. This process may be enhanced somewhat by placing drag-producing devices beneath the buoyant portion of the bottle so that the way the bottle moves will be determined by the water motion a meter or so below the surface.

This principle is used when either dye or a radioactive tracer is placed in the water. Over a period of time the concentration of this pollutant is measured, or the point of its maximum concentration is charted as the tracer moves and disperses through the oceanic volume. In this manner, not only can trajectories be determined, but a great deal can be learned about turbulence from a study of dispersion patterns. This has been a powerful tool in recent years because modern analyzers can measure concentrations of certain fluorescent dyes as low as 1 part in 10^{11}.

With modern electronic equipment it is also possible to track solid objects, such as radio or ultrasonic buoys, placed either at the surface or at some intermediate depth and allowed to drift. Those placed at intermediate depths are usually called *swallow floats,* after John Swallow, who first developed the concept of a neutrally buoyant float. Submerged floats are tracked by means of sonic energy, while those at the surface may be followed more readily by either radar or radio contact.

Another aspect of the Lagrangian concept is based on the fact that a ship drifts and, if it is possible to determine its position with reasonable accuracy, the surface motion of the water may be inferred from the amount and direction of ship drift. This is the most common of all methods for determining surface current because most data for pilot charts are obtained by it.

Data obtained from drift studies are extremely valuable in many practical problems, especially those concerned with fluid transport. Good examples of this are surveys of sedimentation, erosion, and silt transport. A commonly used drifter is a small plastic object that looks very much like a flower, has a small amount of negative buoyancy, and stays close to the bottom. When large quantities of these objects are released, they provide valuable information with respect to motion at the mud-water interface.

611. Eulerian Direct Methods

The most common direct method of measuring current is by means of a meter that is fastened securely at one point in space and that receives its excitation from the passing fluid. These Eulerian types are many and varied and are being used extensively today even within the deep ocean. Satellite naviga-

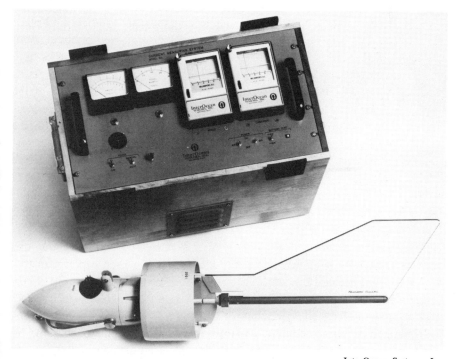

InterOcean Systems, Inc.

Figure 6-2. A propeller current meter.

tion is making it easier and easier accurately to estimate ship drift so that the motion of the water with respect to the earth's surface may be more closely approximated. Generally speaking, Eulerian devices fall into two classes; the dynamic type, which has a rotating sensor, such as a propeller or rotor; and the static type, which has no moving parts. Since measuring very small currents involves the concept of inertia, devices that have very small threshold velocities and time constants and have no rotating members are preferable for such measurements. At the upper end of the scale, where large currents are involved, there seems to be no preference. No means has yet been devised for measuring extremely large currents, such as turbidity currents, which are usually intermittent; what normally happens is that the entire device is carried away by the force of the water.

The first true current meter satisfactorily used at sea was introduced by V. W. Ekman in 1905. Although it is not often used any more, it is interesting to look at its construction since it provided an ingenious solution to a rather knotty problem. It consisted of a propeller connected, by means of a gearing arrangement, to a tube which allowed a small bronze ball to fall into a container every time one hundred revolutions of the propeller had been accomplished. In addition, the container into which the bronze balls fell was compartmentalized and magnetized so that it oriented itself toward magnetic north. Current magnitude could be determined by simply counting the number of balls dropped in a given time period, and direction could be determined by noting which compartment contained the most balls. The entire unit, with its propeller locked, was lowered over the side. When it was desired to start measuring the current, a messenger was dropped, unlocking the propeller. After a given period of time, another messenger was dropped to lock the propeller, and the measurement ended. The total time of operation was simply equal to the time between the dropping of the first and second messengers.

Propellers have long been the favorite rotating sensors of those who design current meters because their characteristics are fairly well known. However, in recent years S-type rotors, typified by the Savonius rotor, have been tending to take their place (*see* Figure 6-4). The advantage of this type of rotor over the propeller is that its response is highly directional and, consequently, vertical motions have a negligible effect on a horizontally oriented meter.

Most modern rotating-sensor current meters count the number of revolutions per unit time and display this count as a measure of the current speed. Four of the more popular methods for producing a countable output are reed switches, magnetic coupling, mechanical counters, and light pulses. Of these, the rotating device that requires the smallest amount of torque is the one that makes or breaks a light beam. The other types require a substantial amount

MEASUREMENT OF FLUID MOTION 71

Hydro Products
A Division of Dillingham Corporation

Figure 6-3. A Savonius rotor current meter.

of mechanical energy to produce a pulse, so that in all cases their threshold velocity is limited by the method by which the pulses are produced.

A reed switch is a small encapsulated switch which is caused to open or close by the proximity of a relatively small magnetic field. In practice it is placed inside a watertight case and a small magnet is attached to the rotor. Every time the magnet passes by the switch a circuit is either made or broken and a pulse is produced.

Magnetic pulses are usually produced by a rotating magnet but, instead of a reed switch, there is a coil placed next to the plane of the rotating magnet. Every time the magnet crosses the field of the coil, lines of force are cut and a voltage is induced in the coil.

Mechanical counters may be geared directly to a digital counter.

Light pulses are produced by some sort of a shutter attached to the rotor breaking a light beam. In this manner, every time a light beam is broken, a photoreceptor produces an electric pulse.

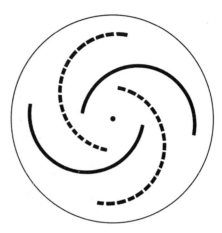

Figure 6-4. Top view of a Savonius rotor.

612. Signal Conditioners

Pulses may be recorded in a number of different ways. They can be recorded directly on magnetic tape or on punched-paper tape as a digital output. They can also be recorded on film as a coded output that can be converted to punched-paper or magnetic tape, or they can be recorded as an analog output on an analog recorder. Another method is to have small dials or counters read current speed and direction directly. Pictures of these dials may be taken at predetermined intervals by a programmed movie camera, and this results in another form of digital output.

If the data are to be used in some sort of computer program, it is obviously desirable to have the type of readout that is suitable for computer use. Figure 6-5 shows a record of current direction and speed obtained from a current meter and recorded on 16-mm. film. This is a coded recording that can be converted into magnetic or punched-paper tape for input to a computer by means of a specially designed converter. Coded pulses are obtained from slotted discs such as the one illustrated in Figure 6-6. The discs are placed over a number of light pipes in such a way that the position of a disc that rotates with compass direction relative to a stationary disc attached to the vane determines the direction of the current. For the current speed, a mark is placed on the film for each pulse in one column, while every tenth pulse puts a mark in another column. If the rotor rotates so rapidly that the dots are too close together to be resolved, the dot indicating 10 revolutions may be

MEASUREMENT OF FLUID MOTION

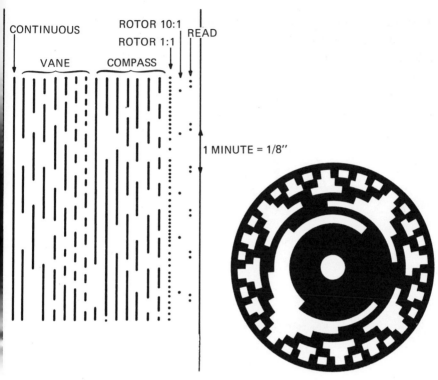

Figure 6-5. Current direction and speed recorded on 16-mm. film.

Figure 6-6. The slotted disc used to produce record shown in Figure 6-5.

used to determine more closely the number of revolutions per unit time. In this way, a large range in speeds may be read.

613. Static Current Meters

There are many types of non-rotating-sensor current meters and we shall look at only a few. These include the pressure-plate type, wherein the ram pressure on a plate is measured; the arrested-rotor type, which measures the torque necessary to constrain the rotor motion; and the pressure, or pitot-tube, type, where the difference in pressure at two different points in the flow field is determined. All these have been used, but none is presently popular for measuring oceanic currents.

A simple device that requires no moving parts and can be used from anchored vessels or platforms is the captured current drag, as described by D. W. Pritchard and W. V. Burt. It consists of a tethered hydrodynamic plate which is immersed in the moving fluid. The motion of the fluid tends to push the plate in the direction of flow, while at the same time the weight of the plate tends to keep it vertical. An equilibrium position is reached when the weight and the drag force on the plate are in balance, producing some angle in the tether line (*see* Figure 6-7). This angle is a measure of the speed.

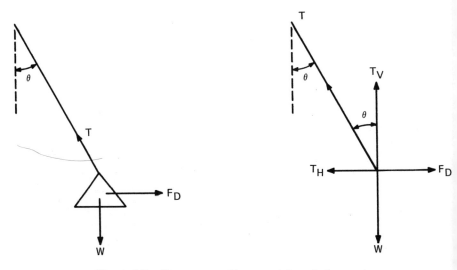

Figure 6-7. Forces present in current-drag deployment.

For Reynolds numbers over a value of about 1,000, the drag coefficient of a plate is a constant, and that is why this hydrodynamic configuration was chosen. Simple trigonometry shows that the horizontal component of the tension force in the cable must be equal to the drag force exerted on the plate

$$T_H = F_D \tag{7}$$

or, expressed in terms of other parameters,

$$\left(\frac{W}{\cos \theta}\right) \sin \theta = Kv^2 \tag{8}$$

since, from Figure 6-7

$$T_H = T \sin \theta$$
$$T_V = T \cos \theta = W$$
$$T = \frac{W}{\cos \theta} \tag{9}$$

Also

$$F_D = C_D \frac{A\rho v^2}{2} = Kv^2 \tag{10}$$

where T = tension in the cable (T_H is the horizontal component, T_V is the vertical component)
θ = angle the cable makes with the vertical
W = weight of the plate in water

v = speed of the fluid
F_D = drag force
C_D = drag coefficient
A = area of plate
ρ = density of the fluid.

Therefore

$$v^2 = \frac{W}{K} \tan \theta \tag{11}$$

The plate configuration actually used is in the form of a cross made by placing two pieces of plywood at right angles to each other in egg-crate fashion (*see* Figure 6-8). This captured drag makes for a rather simple device, since all that is necessary is to lower it to the desired depth and measure the angle of the cable to the vertical to determine the existing current. The readings obtained are somewhat crude, but for measuring current in estuarine and shallow coastal water, this device is as good as any.

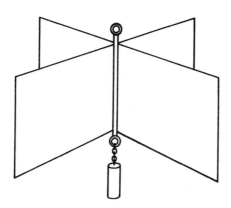

Figure 6-8. A simple current drag.

This general concept of the change of the angle with drag force is also used, in an inverted sense, as a bottom-current measuring device wherein a buoyant tube is lowered and the angle that it makes with the vertical is a function of the current.

Another non-rotating current sensor is the electromagnetic one. In principle, it is similar to the GEK in that it measures the EMF generated by a conductor moving in a magnetic field. Instead of measuring the electrical gradient of the ocean, however, the potential between two closely spaced electrodes in a high-density, artificially-generated magnetic field is monitored. Thus, signals large enough to be easily handled are generated by a sensor the same size as the other common current meters.

614. Turbulence Measurement

None of the devices thus far described are adequate for measuring small-scale turbulence. To meet this need two classes of devices have been suggested: one uses an overheated wire or semi-conductive type of material, while the other uses the passage of ultrasonic energy through the fluid medium.

An overheated wire (called the hot-wire anemometer when used in the atmosphere) operates on the principle that a body at temperature above the surround conducts heat to the surround at a rate determined by the renewal of cold fluid around the heated body. In other words, the faster the fluid moves by, the more heat is lost by the body. This loss in heat can be measured either by determining how much additional heat energy is required to keep the temperature constant, or by simply determining the change in temperature of the body as it is placed into the flow. In either case the heat inertia and the physical size of these devices are small enough to allow them to be used for measuring small eddies in the natural environment. One disadvantage of the devices is that they give no indication of current direction: they give only a measure of current magnitude. Another complication is that temperature compensation is necessary because changes in ambient temperature will also affect the rate at which heat is lost by the overheated body. Generally speaking, hot-wire current meters are still laboratory devices and have not been used in the field for routine current measurements.

Sonic devices can generally be divided into two groups. In one group, the speed of sound in the direction of the current is compared with the speed of sound in the opposite direction. Since a moving fluid carries sound energy along with it, there will be a difference related to the speed of motion of the fluid. The other group is based on the fact that any naturally occuring fluid contains small inhomogeneities from which it is possible to reflect sound energy: sound energy is sent to some point in the water from which energy is reflected; a receiver is pointed at that point and it picks up the reflected energy. If the parcels that do the reflecting are in motion, there will be a doppler effect and the frequency of the received signal will then differ from that which is transmitted. This frequency shift is proportional to the water motion in the direction of the sound propagation. The major advantages of this type of device are that it can measure a very small volume and it has a very rapid response. It also samples at a rapid rate, so that eddies less than a centimeter in size with periods on the order of milliseconds can be examined. Both of these sonic devices can be mounted in arrays of three at right angles to each other, so that the total vector current velocity can be obtained directly.

Another type of fluid-motion sensor that holds a great deal of promise uses a phenomenon known as the *von Karman vortex street,* or the *singing wire* (*see* Figure 6-9). A well-known effect in hydrodynamics is seen when

Figure 6-9. A von Karman vortex street.

a cylinder placed in a moving fluid causes a wake of small vortices to be built up behind the cylinder in a line called a *vortex street*. The rate at which these vortices form and decay is a rather simple function of the speed of the fluid and the diameter of the cylinder, as given in

$$f = \frac{Kv}{D} \tag{12}$$

where f = frequency of induced oscillation in the wire
 D = diameter of the wire
 K = empirically determined constant related to Reynolds number
 v = speed of fluid.

Thus, with proper choice of wire dimensions and material, it is possible to design a device that will oscillate within a reasonable range of frequencies for the measured speeds desired. An alternative method is to measure the rate at which vortices are being shed by sampling the fluid immediately downstream from the wire.

615. Satellite Detection of Currents

If current systems are composed of waters whose characteristics are slightly different from those of the surrounding water, which is either not in motion or is in motion in another direction, it is possible to locate these currents by means of satellite sensors. The most common method is by using visible light, both wide-band and narrow-band. Many currents are composed of water that has a higher or lower silt content than the surrounding water, and they show up very plainly on a satellite photograph.

In addition to this, infrared data can also be used to locate any temperature differences that may be present between different water masses. This is true particularly in a current such as the Gulf Stream, whose temperature is somewhat higher than that of the water on either side. Infrared photographs can very easily locate the Gulf Stream at a particular instant in time.

616. Current Units

Units used to describe oceanic currents are involved with two lengths. One is the nautical mile and the other is the meter. A nautical mile is equal to one minute of arc at the equator; therefore, a 90° arc around the equator is equal to 5,400 nautical miles. A meter, on the other hand, is defined in terms of the distance along a meridian of longitude in such a manner that 10^7 meters is equal to 90° of arc from the north pole to the equator. Assuming the earth is a sphere, this means that 1 kilometer is equal to about 0.54 nautical miles, or 1 nautical mile is equal to about 1.86 kilometers. A nautical mile per hour, or a knot, equals 0.515 meters per second, and one meter per second is equivalent to a speed of 1.94 knots.

617. Direction Determination

Current direction is always specified in the direction toward which the flow is going. Mariners are interested in where their ships will end up, so they are concerned with the direction toward which the flow is moving rather than the direction from which it came. Therefore, a current flowing from the northwest to the southeast is called a southeast current.

Current direction is usually measured by comparison between the direction of flow and that of a compass. For this purpose, a vane is allowed to swivel until it lines itself up in the direction of the current, and when this occurs its direction is compared to a compass reading.

Another way of determining current direction is to use two or more directional current meters placed at right angles to each other. These may be channelized rotating devices, such as propellers within a tube, or they may be sonic-type devices. If the orientation of the current meters is known, the true current vector can be obtained.

Although simple in theory, current vanes are less than perfect in practice, because the time constants normally associated with them are rather large. Often a vane will "hunt," and when the currents are small it will occasionally not move at all. With small-magnitude currents, a vane reacts sluggishly, and with rapidly changing currents, it is hard-pressed to keep up with the changes.

618. Vector Averaging

Once current magnitude and direction have been obtained, there is again the problem of averaging. If the magnitude and direction are averaged separately, the answer obtained will be quite a bit different from that obtained by vector average of the instantaneous current. Thus, a current of 1 knot to the south averaged with a current of 1 knot to the north gives a zero current: a meaningless answer. Determining a vector average is a problem all by itself, but it has to be solved if a meaningful result is desired. For this reason it is general practice for the current direction and current magnitude data to be kept separate and averaged or analyzed independently.

619. State of the Art

More new current sensors have been introduced in recent years than at any time in the history of oceanographic instrumentation. These new devices are especially important, not only because their threshold velocities are very low, but also because their frequency responses are broader than those of any previous units. The problem of measuring deep-ocean currents remains unsolved, however, and there is nothing on the horizon that will materially affect this situation.

Sources and Additional Reading

Babiy, V. I. "Some Problems of Measuring Current Velocity by the Doppler Method." (In Russian). In: *Metody i Pribory dlya Issledovaniya Fizicheskikh Protesessov v Okeane.* A. N. Paramonov, ed., Izd-vo Naukova Dumka, Kiev. Translation: JPRS: 39, 13 Feb. 1967 (Clearinghouse Fed. Sci. Tech. Info., U. S. Dept. Commerce).

Beyer, F., E. Foyn, J. T. Ruud, and E. Totland. "Stratified Currents Measured in the Oslofjord by Means of a New Continuous Depth-Current Recorder, the Bathyrheograph." *J. Cons. Perm. Int. Explor. Mer,* 31(1), 1967.

Cannon, G. A., and D. W. Pritchard. "A Biaxial Propeller Current-Meter System for Fixed-Mount Applications." *J. Mar. Res.,* 29(2), 1971.

Carruthers, J. N. "Various Desiderata in Current-Measuring and a New Instrument to Meet Some of Them." In *Studies on Oceanography dedicated to Professor Hidaka in Commemoration of his Sixtieth Birthday,* 1964.

———. "The Plastic Seabed 'Oyster' for Measuring Bottom Currents." *Fisk Dir. Skr. Ser. HavUnders,* 15(3), 1969.

Chalupnik, J. D., and Pl. J. Green. "A Doppler Shift Ocean Current Meter." *Mar. Sci. Instrum.,* 1, 1962.

Dextraze, R. E. "A Three-Axis Ducted Impeller Current Meter System." *Mar. Sci. Instrum.,* 4, 1968.

Dietrich, G., and G. Siedler. "Ein neurer Dauerstrommesser" (English abstract). *Kiel. Meeresforch,* 19(1), 1963.

Duncan, C. P. "Disadvantages of the Olson Drift Card, and Description of a Newly Designed Card." *J. Mar. Res.,* 23(3), 1965.

Eagelson, P. S., and W. P. M. Van De Watering. *A Thermistor Probe for Measuring Particle Orbital Speed in Water Waves.* U. S. Army Coast. Engng. Res. Center, Techn. Mem. No. 3, 1964.

Forstner, H., and K. Rutzler. "Two Temperature-Compensated Thermistor Current Meters for Use in Marine Ecology." *J. Mar. Res.,* 27(2), 1969.

Frasseto, R. "A Neutrally-Buoyant, Continuously Self-Recording Ocean Current Meter for Use in Compact, Deep-Moored Systems." *Deep-Sea Res.,* 14(2), 1967.

Fukuda, M. "The Spherical Current Meter." *J. Oceanogr. Soc.,* Japan, 21(3), 1965.

Gakkel, J. J., and L. P. Samsonia. "The First Drifting Radio Buoys." (In Russian). *Okeanologia, Akad. Nauk. SSSR,* 1(4), 1961.

Gaul, R. D., J. M. Snodgrass, and D. J. Cretzler. "Some Dynamical Properties of the Savonius Rotor Current Meter." *Mar. Sci. Instrum.,* 2, 1963.

Holmes, J. F. "Wide-Range Flow Meter for Oceanographic Measurements." *Symposium, Ocean Sci. and Ocean Eng.* Mar. Techn. Soc., Am. Soc. Limnol. Oceanogr., 2, 1965.

Horrer, P. L. "Methods and Devices for Measuring Currents." In *Estuaries,* G. H. Lauff, ed. Am. Ass. Advmt Sci., 83, 1967.

Iwata, E., S. Ohta, and T. Suguro. "Some Experiments in Geomagnetic Electrokinetograph (GEK)—I." (In Japanese; English abstract). *Bull. Tokai Reg. Fish. Res. Lab.,* 56, 1968.

Kestner, A. P. "Electromagnetic Velocity Meter of the Turbulent Water Flow-EMK-1. Questions of Physical Oceanology." (In Russian; English summary). *Trudy Inst. Okeanol., Akad. Nauk, SSSR,* 66, 1963.

Kolenikov, A. G., N. A. Panteleiv, V. D. Pisarev, and P. V. Vakylov. "An Abyssal Autonomic Turbulence Meter, an Instrument for the Registration of Turbulent Fluctuations of Velocity and Temperature in the Ocean." (In Russian). *Okeanologiia, Akad. Nauk, SSSR,* 3(5), 1963. Abstract in *Soviet Bloc Res. Geophys., Astron. and Space,* No. 74.

Kronegold, M., and W. Vlasak. "A Doppler Current Meter." *Mar. Sci. Instrum.,* 3, 1965.

Kurdiavtsev, N. F., and E. G. Nikiforov. "On the Selection of Rational Constructions of Instruments and the Estimation of their Optimum Parameters Intended for Current Measuremtnts in a Layer Enveloped by Wave Activity." (In Russian). *Okeanol. Akad. Nauk, SSSR,* 4(3), 1964.

Kvinge, T. "An Acoustic Current Meter." *Arbok Univ. Bergen, Mat. Naturv Ser.,* 7, 1965.

Lacroix, P., and L. Laubier. "Description étalonnages et essais preliminaires d'un nouveau courantometre enregistreur." *Cah. Oceanogr., CCOEC,* 16(9), 1964.

Lee, A. J., D. F. Bumpus, and L. M. Lauzier. "The Sea-Bed Drifter." *Internat. Comm. Northwest Atlantic Fish. Res. Bull. No. 2,* 1965.

Lester, R. A. "High-Accuracy, Self-Calibrating Acoustic Flow Meters." *Mar. Sci. Instrum.,* 1, 1962.

Martin, J. "Expérience de G.E.K. vertical dans le Détroit de Gibraltar." *Cah. Océanogr., CCOEC,* 16(5), 1964.

Michelena, E., F. C. Elder, and H. K. Soo. "A Triaxial Flowmeter for Wave Motion Measurements." *Proc. 11th Conf. Great Lakes Res., 1968.*

Nakamura, S. "A Study on Photoelectric Current Meters." *Bull. Disast. Prev. Res. Inst.*, Kyoto Univ., 15(1), 1965.

Naumenko, M. F., V. T. Paka, M. A. Strunina, B. F. Trinchuk, and K. I. Chigrakoy. "Apparatus and Methods of Investigation of Some Kinds of Turbulent Mixing." (In Russian). *Mat. Vtorai Konf. Vzaimod. Atmos. Gidrosfer, v Severn. Atlant. Okean., Mezhd. Geofiz. God.*, Leningr., 1964.

Neiman, V. G. "On the Measurement of Currents from a Drifting Ship." (In Russian). *Okeanologiia, Akad. Nauk, SSSR*, 6(1), 1966.

Niskin, S. J. "A Low-Cost Bottom Current Velocity and Direction Recorder." *Mar. Sci. Instrum.*, 3, 1965.

Ortolan, G. "New Current Meter for Measuring Currents Near the Bottom." *Int. Hydrogr. Rev.*, 43(2), 1966.

Pikush, N. V. "The Outflow Method and Tachymeter for the Water Current Speed Measuring." (In Russian). *Met. i Gidrol.*, 2, 1965.

Pochapsky, T. E. "Measurement of Small-Scale Oceanic Motions with Neutrally-Buoyant Floats." *Tellus*, 15(4), 1963.

Popov, I. K. "Use of the GEK in Oceanological Investigations by the Thirteenth Soviet Antarctic Expedition." (In Russian). *Inform. Bull. Sovetsk. Antarkt. Exped.*, 71, 1968. Transl.: Scripta Tecnica. Inc. for A.G.U., 7(3).

Resch, F. J., and J. D. Irish. "Quartz Crystals as Multipurpose Oceanographic Sensors—II. Speed." *Deep-Sea Res.*, 19(2), 1972.

Richardson, W. S., A. R. Carr, and H. J. White. "Description of a Freely Dropped Instrument for Measuring Current Velocity." *J. Mar. Res.*, 27(1), 1969.

Rychkov, V. S., and A. N. Kodarev. "An Acoustic Current Meter." (In Russian). *Okeanologiia*, 11(4), 1971.

Sasaki, T., S. Watanabe, and G. Oshiba. "New Current Meters for Great Depths." *Deep-Sea Res.*, 12(6), 1965.

Solov'yev, L. G. "On Measurements of Electric Fields in the Seas." (In Russian). *Dokl. Akad. Nauk, SSSR*, 138(2), 1961.

Squier, E. D. "A Doppler Shift Flowmeter." *Mar. Sci. Instrum.*, 4, 1968.

Stass, I. I. "Automatic Photoelectric Deep-Water Instrument for Recording Ocean Current Elements." (In Russian). *Trudy Morsk. Gidrofiz. Inst.*, 26, 1962. Transl.: Scripta Tecnica, 26, 1964.

Talbot, G. B. "Drift Bottle Modification for Air Drops." *Trans. Amer. Fish. Soc.*, 93(2), 1964.

Thomas, R. S. "Measurement of Ocean Current Shear Using Acoustic Backscattering from the Volume." *Deep-Sea Res.*, 18(1), 1971.

Thorndike, E. M. "A Suspended-Drop Current Meter." *Deep-Sea Res.*, 10(3), 1963.

Tungate, D. S., and D. Mummery. "An Inexpensive Mechanical Digital Flowmeter." *J. Cons' Perm. Int. Explor. Mer.*, 30(1), 1965.

von Arx, W. S. *An Introduction to Physical Oceanography.* Addison-Wesley Publishing Co., Boston, 1962.

———. "Measurements of Subsurface Currents by Submarine." *Deep-Sea Res.*, 10(3), 1963.

Webb, D. "Measurement of Vertical Motion in the Ocean." *Woods Hole Oceanographic Institution,* 1965.

CHAPTER 7

Light-Associated Measurements

701. Light Losses

Since man's eyes form the primary link between him and his environment, it is not too surprising that he should be interested in the measurement of light losses within the oceanic regime. This interest is not lessened even by the fact that the losses are generally quite large. For example, light may have losses 10,000 times as great as those of sound within the hydrosphere.

The type of loss that is of interest in most optical systems is involved with two physical processes, absorption and scattering. These processes occur in varying amounts relative to one another, and are related to some of the characteristics of the ocean, along with the wavelength of the light being used. To begin, let us define these two terms in order to more exactly specify our problem.

702. Absorption and Scattering

Absorption is the conversion of light energy to heat energy, while *scattering* is the redirection of light energy by the presence of scatterers. Scattering may be treated as simple specular reflectance, refraction and diffraction through and around the particles, or even by absorption and reradiation of the electromagnetic energy. These three tacks have been taken with some success, depending upon the relative particle size, but only scattering in terms of its redirection of light energy is considered here.

A rather simple mathematical relationship describes the diminution of light energy as it passes through the water, if the water path may be considered to be homogeneous (*see* Figure 7-1). This relationship is called Lambert's law and is expressed as

$$\frac{I_l}{I_0} = e^{-al} \tag{1}$$

where I_0 = initial light intensity
I_l = light intensity after passing through a path l units long
a = absorption coefficient expressed as a reciprocal length.

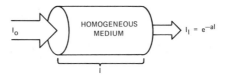

Figure 7-1. Lambert's law.

Equation (1) describes the absorption process, but similar relationships may be used to describe any type of loss. Equation (2) describes scattering loss, and Equation (3) describes some combination of absorption and scattering called *attenuation*. Note that the only difference between the three equations is the specification of the loss coefficient.

$$\frac{I_l}{I_0} = e^{-bl} \qquad (2)$$

$$\frac{I_l}{I_0} = e^{-cl} \qquad (3)$$

where b = scattering coefficient
c = attenuation coefficient.

If the attenuation described by Equation (3) is that experienced by a beam, then

$$c = a + b \qquad (4)$$

Oftentimes the reciprocal of the loss coefficient, the characteristic length $\mathcal{L}$, is used to describe the medium, because a length has a greater intuitive meaning than a reciprocal length. Thus

$$\mathcal{L}_a = 1/a, \; \mathcal{L}_b = 1/b, \; \mathcal{L}_c = 1/c \qquad (5)$$

or from Equations (1), (2), and (3) it can be seen that the characteristic length is also the length through which light must travel to be diminished to $1/e$ of its initial value. Note that even though the total attenuation coefficient is equal to the sum of the absorption and scattering coefficients [Equation (4)] the characteristic attenuation length is not equal to the sum of the characteristic absorption and scattering lengths since

$$\mathcal{L}_c = \frac{1}{c} = \frac{1}{a+b} = \frac{1}{\frac{1}{\mathcal{L}_a} + \frac{1}{\mathcal{L}_b}} = \frac{\mathcal{L}_a \mathcal{L}_b}{\mathcal{L}_a + \mathcal{L}_b} \qquad (6)$$

The characteristic length is similar in concept to the time constant (*see* Section 204).

As light passes through water it is subject to the effects of both absorption and scattering. However, in many cases, scattering is so small that absorption is the primary cause of light attenuation. The spectral absorption characteristics of water are plotted in Figure 7-2, and it is apparent that water is a selective absorber. Note that both the far red and the far violet ends of the visible spectrum are attenuated very markedly by absorption, while in the blue-green portion, peaked at about 470 nanometers, the characteristic length has been measured to be more than 100 meters. This length is quite small compared to similar measurements of audible sonic energy, but compared to the rest of the electromagnetic spectrum it represents a definite transmission window. This filtering action of water, which allows the blues and greens to pass relatively easily but tends to prevent the passage of reds, is the major reason why colors appear so distorted to a diver who has no artificial illumination.

The absorption properties of water can be very easily specified as was done above, but scattering is another story. Due to the fact that scattering is primarily a function of the suspended materials in the water, the amount and type of scattering varies with the amount and type of suspensoids. It would be expected, for example, that scattering effects for small clay particles and those for similar sized plankton would be different, as would those for small and large particles of the same material. Not only does scattering vary both in magnitude and direction with the amount of particulate matter in suspension, but it varies with the size of the particles relative to the wavelength of impinging light and the index of refraction of the scattering material.

Generally speaking, particles that are smaller than the wavelength of light are much more effective in scattering blue light than red, the extreme case described by Rayleigh scattering. However, as the particle size increases the amount of red light scattered also increases so that some relative particle size compared to the light wavelength is reached where both red and blue

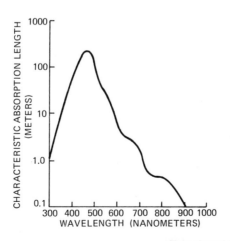

From *Optical Properties of the Sea*

Figure 7-2. Spectral absorption properties of water. (After James and Birge, 1938)

light are scattered equally. As the particle size is further increased with respect to the light wavelength, we find that more and more red light is scattered, and finally the whole thing repeats itself as the amount of red light scattered decreases with a further increase in relative particle size. This is illustrated in Figure 7-3.

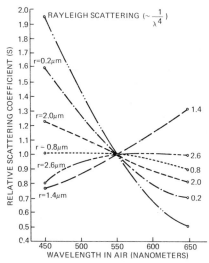

From *Optical Properties of the Sea*

Figure 7-3. Relative scattering by clay particles of different sizes when suspended in water.

Thus it would appear that, for both absorption and scattering measurements, not only must the amount of light lost be measured but it must be known what wavelength of light is being used in the measurement. Since both absorption and scattering are so markedly related to the wavelength of light, unless that parameter is specified, the data are meaningless.

703. Irradiance and Beam Transmittance

Two of the most frequently measured light-connected parameters in the ocean are irradiance and beam transmittance. Although both of these parameters involve the physical processes of absorption and scattering, their relationship is quite different and therefore the devices used to measure them are also markedly different. *Irradiance* is defined as the light flux falling on a horizontal unit area, while *beam transmittance* is defined in terms of the light flux contained within a beam which successfully transits a unit pathlength.

The implication in the definition of beam transmittance is that any light that once leaves the beam, that is, changes its travel direction, is counted as being lost from the beam. Irradiance, on the other hand, implies that for each flux line that strikes a horizontal surface the angle with the surface must be considered. A given value of flux striking a unit area at one angle would not produce the same value of irradiance as would the same amount of flux striking the same area at some other angle, because the amount of area covered by the flux would not be the same. This is shown in Figure 7-4.

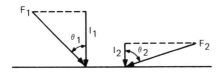

Figure 7-4. Flux, F_1, striking a horizontal surface at one angle, θ_1, results in irradiance value I_1. The same flux, F_2, striking the same surface at angle θ_2 results in irradiance value I_2.

It follows then that a receiver for measuring irradiance must have an acceptance angle of 180°, while a receiver for measuring beam transmittance must have a theoretical acceptance angle of 0°. The ideal receivers for these two measurements are the cosine collector and the Gershun tube, respectively.

704. The Cosine Collector

A cosine collector is a device that measures the amount of flux that strikes a horizontal surface and multiplies each flux component by the cosine of the angle the line of flux makes with the vertical. A simple method of approaching this ideal (one can only "approach" this state because there is no such thing as a perfect cosine collector) is to use opal glass, which transmits only an amount of light roughly proportional to the cosine of the angle made by incident light. In this manner, light coming in normal to the glass surface causes a maximum response from a photosensitive receptor placed beneath the glass, while a cosine proportional response is produced from any other angle.

705. The Gershun Tube

The Gershun tube, on the other hand, minimizes the angle of acceptance. It is a long, thin tube, blackened on the inside, so that only light traveling in a direction essentially parallel to the tube can pass through, the rest being absorbed by the blackened walls of the tube. The acceptance angle is obviously a function of the length and diameter of the tube, and for small angles the tangent of the acceptance angle is simply the ratio of tube diameter to tube length. We can see that, as this ratio is made smaller and smaller, the acceptance angle decreases, and, in the limit, it is possible to approach an acceptance angle of zero with an infinitely long tube whose diameter is infinitely small.

In practice, the Gershun tube is usually approximated by a lens-aperture arrangement, as illustrated in Figure 7-5. In the case illustrated, rays of light entering the lens at greater than the acceptance angle are focused outside the aperture while, within a certain acceptance angle, they pass through the hole. The acceptance angle γ of this lens-aperture system can be described simply by

$$\gamma/2 = \tan^{-1} y/f \qquad (7)$$

where y = aperture radius
f = lens focal length.

For very small angles, the acceptance angle in radians is given by

$$\gamma \simeq d/f \qquad (8)$$

where d is the aperture diameter.

The acceptance angle may thus be varied by changing either the lens or the size of the aperture.

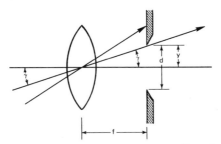

From *Optical Properties of the Sea*

Figure 7-5. A lens-aperture Gershun tube.

706. Wavelength Specification

One aspect of light-loss measuring devices common to all is that involved with the specification of the wavelength of light used in the measurement. Many light-loss measuring instruments do not take this into account and consequently produce misleading data. Let us look at a typical example of what might happen with a device that had a broad-bandwidth receiver.

A beam transmittance meter consists essentially of a light source, a light path through the water, and a light receiver, such as a photocell. Figure 7-6 shows the light-response characteristics of a typical incandescent light source, a water path of one meter, and a photocell of the resistive type. If these three are placed together as a system, the response of the system as a function of wavelength may be seen from Figure 7-7. Let us now change the pathlength involved so that only one-tenth of a meter, or 10 centimeters, of water is interposed between the light source and the light receiver. Figure 7-7 shows the response of this system with only a 10-centimeter pathlength. Note that it is markedly different in the red end of the spectrum.

When the total (all wavelengths included) attenuation coefficient, as seen by this variable-pathlength system, is plotted as a function of pathlength, the

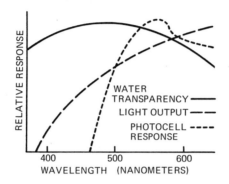

Figure 7-6. Wavelength characteristics of water, light, and photocell.

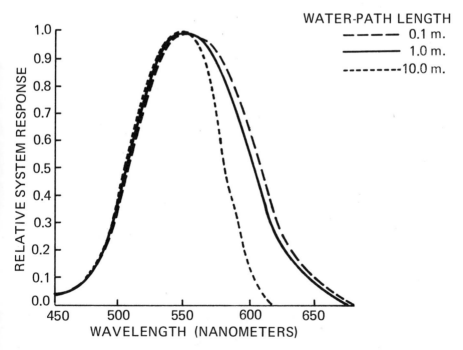

Figure 7-7. System response for pathlengths of ten centimeters, one meter, and ten meters.

measured attenuation coefficient changes markedly as the pathlength changes (*see* Figure 7-8). This is simply because the light composition changes as it moves through greater thicknesses of water. Initially a large portion of the source's energy was contained in the red end of the spectrum, but these wave-

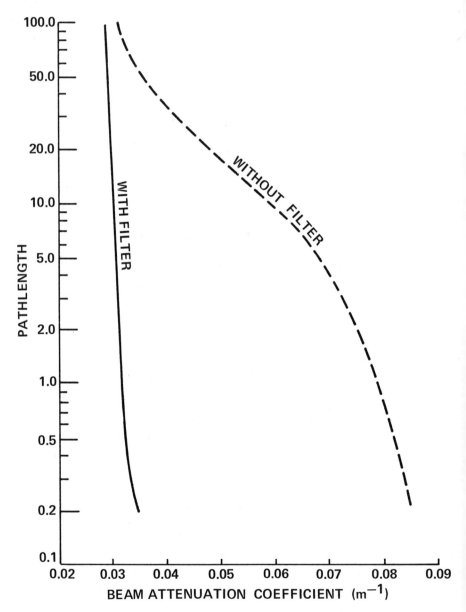

Figure 7-8. Variation of the apparent beam attenuation coefficient with pathlength for a filterless system and for a system containing a filter.

lengths were lost in a short distance. Thus, for short distances the total attenuation coefficient was high. For larger pathlengths however, the relative amount of light lost was smaller, because the remaining wavelengths were not attenuated as strongly.

On the other hand, the use of a color filter, such as a Wratten 61, has the effect of eliminating from any consideration all wavelengths outside the filter's range. In Figure 7-9 the transmittance of the Wratten filter is plotted as a function of wavelength, as is the relative system response when the filter is included. This response can be compared to that without the filter shown in Figure 7-7. In Figure 7-8 a plot of the variation of attenuation coefficient with pathlength shows the marked improvement achieved with the filter in the system. Thus, unless some method of delineating a small band of wavelengths within the optical system is used, the measured attenuation coefficient is directly related to the distance between the light and the photocell. No two instruments whose pathlengths differ give directly comparable measurements.

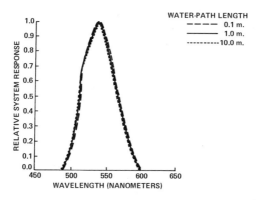

Figure 7-9. Relative response of a light-water-photocell system with a Wratten 61 filter for three different water pathlengths. System response is essentially constant for all pathlengths.

707. The Immersion Effect

Another problem involved in instruments to be used within a water environment is that of immersion effect. Due to the fact that the amount of light reflected from any interface is a function of the refractive indices of the two materials involved, the amount of light lost at a window surface with a water path is different from that lost at a window surface with an air path. There is thus a difference when the device utilized is an irradiance meter in terms of the angle the flux makes with the receiving surface. All this is summed up in Fresnel's equation

$$\rho = \frac{1}{2}\left(\frac{\tan^2(i-r)}{\tan^2(i+r)} + \frac{\sin^2(i-r)}{\sin^2(i+r)}\right) \tag{9}$$

where ρ = reflectance of the interface
i = angle to the normal made by the incident light
r = angle to the normal made by the refracted light.

For normal incidence a glass-air interface reflects about 4% of the incident light, while a glass-water interface reflects about 2% of it. Thus an instrument calibrated in air will be somewhat in error when immersed in water.

From Equation (9) it can be seen that this effect is more and more pronounced as the angle from the vertical changes. Both water and glass are dispersive media, that is, their index of refraction varies as the wavelength of the light, so it can be expected that the immersion effect will also be a function of the wavelength.

708. Relative Irradiance

In order to examine the measurement of relative irradiance, let us consider the irradiance at a depth d and the irradiance at some greater depth d'. These depths can, if desired, be enclosed within a hypothetical column, as shown in Figure 7-10. The energy passing through a horizontal plane constrained by this column at depth d has an irradiance equal to e. The irradiance at depth d' is, then, equal to e', which is somewhat smaller than e. This is so because some of the flux that leaves the plane at d is scattered backwards and lost, some is scattered through the sides of the column, and some reaches d' by a more or less direct path.

Since the ocean is relatively large in a horizontal direction, it can be imagined that there are many other columns, besides the one we have initially drawn, with similar light paths. Therefore, some light from the others is entering the sides of the original column. If we allow an ocean of infinite horizontal expanse, we can probably assume with a fair degree of accuracy

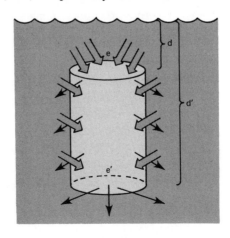

Figure 7-10. **A hypothetical irradiance column.**

that the light lost through the sides of our column is replenished through these same sides from the adjacent sea. Therefore, with the exception of the backscattering, the major effect of scattering during the process of irradiance diminution is simply an increase in effective pathlength. Light is not so much lost as it is caused to travel a greater distance and therefore be subject to greater absorption.

709. Beam Transmittance

Relative irradiance may be contrasted with the type of light diminution during the process of beam transmittance. Here, a light source produces a parallel beam that passes through the water to a receiver that accepts only light traveling in the same direction as it was when it left its source. In other words, the acceptance angle of the receiver is vanishingly small. Any light scattered out of a beam never reaches the receiver because of its small acceptance angle; consequently, in the process of beam transmittance, the only light received is that which either has been scattered in a forward direction or has not been scattered at all (*see* Figure 7-11). Although beam transmittance is obviously a measure of both absorption and scattering, in turbid regions where scattering predominates, it is a good measure of scattering alone.

Since the processes of relative irradiance (sometimes called *extinction*) and beam transmittance (sometimes called *attenuation*) are quite different, the type of activity planned will determine which type of light-diminution is of interest. If it is desired to illuminate some underwater object, the process by which the light is diminished as it travels through the water is called *extinction*. This is independent of the type of source used. It is obviously of no import in which direction the light is traveling when it reaches the object, as long as it gets there. However, if it is desired to view the object, it is of extreme importance that any prediction of visual range be made in terms of beam transmittance. An image is formed by light traveling in a straight line from object to eye. Any light scattered into the eye tends only to fuzz an image rather than add to it. Consequently, the measurement of beam trans-

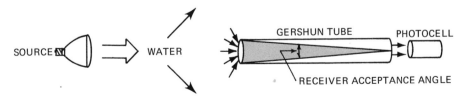

Figure 7-11. A beam transmissometer showing that only light received within the acceptance angle is recorded by the photocell.

mittance is useful in determining possible visibility criteria, whereas the measurement of relative irradiance is useful in determining illumination criteria.

710. Irradiance-Measuring Devices

Irradiance-measuring devices are the simplest of all light-loss measuring instruments. They usually consist of some type of photosensitive component fronted by a combination of filter and opal glass. This instrument is lowered progressively into the ocean and, as this is being done, readings of the irradiance are made at various depths. Relative irradiance is obtained by comparing the irradiance at one level in the ocean with that at another level.

The simplest irradiance-measuring instrument is the Clarke photometer, which consists of a barrier-layer photocell, fronted by a piece of opal glass with provision made for placing color filters in the optical path. This may be used in conjunction with a similar unit turned upside down so that the upwelling light is measured at the same time as the downwelling light.

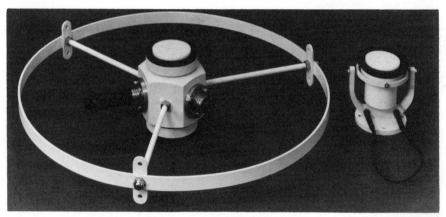

Marine Advisors, Inc.

Figure 7-12. An irradiance meter.

A somewhat more complex instrument for measuring irradiance, called a *spectroradiometer,* has been developed by Scripps Institute of Oceanography and is described by J. E. Tyler and R. C. Smith (Tyler and Smith, 1970). In place of the color filter is a rather intricate system that selects wavelength by a series of gratings. The gratings make it possible to obtain a bandwidth of about 5½ nanometers—ten times better than can be obtained with optical absorption filters. Of course, with a system of this type, the total amount of light getting through is very much lessened, and therefore the photo receptor system has to be much more sensitive than the simple barrier-layer photocell plus microammeter used in the Clarke unit. The Scripps spectroradiometer incorporates a phototube and the associated electronics for amplification. It is suitable for deep-ocean measurements and with it data can be obtained at about 5-nanometer wavelength intervals, allowing a rather detailed profile of irradiance as a function of wavelength to be acquired.

711. Beam-Transmittance Meters

The most commonly used beam-transmittance meter (transmissometer or alpha meter) consists of a light source, a lens system for collimation, a pathlength within the water medium, a lens-aperture system to limit the receiver acceptance angle, and some sort of photoelectric receiver. We have already

Figure 7-13. A beam-transmittance meter.

discussed the instrument incompatibility that arises if some sort of a wavelength-bandwidth limiting device, such as a filter, is not used as a part of the optical system. It can also be shown (J. Williams, 1970) that this conventional type of transmissometer measures its own design parameters just as much as it measures the optical characteristics of the water. When the receiver aperture and the acceptance angle are small enough to be the limiting features of the optical system, the transmittance in terms of instrument parameters may be given by

$$T = \frac{F_L}{F_o} \left(\frac{D + 2L \tan \alpha/2}{D + 2L \tan \beta/2}\right)^2 \left(\frac{g + 2R \tan \alpha/2}{g + 2R \tan \beta/2}\right)^2 \tag{10}$$

where T = transmittance
  F_L = light flux reaching the receiver window with a water path
  F_o = light flux reaching the receiver window with an air path
  D = beam diameter
  L = path length
  α = source spreading angle
  β = spreading of beam due to scattering by the medium
  g = receiver aperture
  R = distance from the receiver aperture to the photocell.

These parameters are depicted in Figure 7-14.

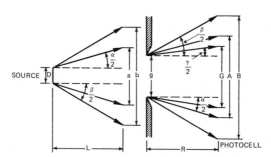

From *Optical Properties of the Sea*

Figure 7-14. Transmissometer parameters. The source is on the left; the receiver on the right.

It may be noted that the light-flux loss is not the only parameter involved in the specification of the transmittance but also the beam diameter, the path length, the source spreading angle, and the receiver optics. Thus it may be seen that two instruments of somewhat different design will give different results in the same water.

The measuring of beam transmittance is really not improved a great deal when certain aspects of the system are made small enough to be neglected.

LIGHT-ASSOCIATED MEASUREMENTS

Even when the ratios of beam diameter to path length, and receiver aperture to aperture-photocell distance are made as small as possible, the transmittance as given by Equation (11) is still dependent on instrument parameters

$$T \simeq \frac{F_L}{F_o} \left(\frac{D}{2L \tan \beta/2}\right)^2 \left(\frac{g}{2R \tan \beta/2}\right)^2 = \frac{F_L}{F_o} \left(\frac{Dg}{4RL \tan^2 \beta/2}\right)^2 \quad (11)$$

Probably, therefore, even if a laser were used as the source in a conventional beam-transmission meter, the resultant transmittance reading would still reflect some of the parameters of the instrument itself.

There appear to be two ways out of this dilemma. One is to standardize beam length, diameter, receiver aperture, etc. for all instruments. If this were done, all would have the same effect on the transparency reading and readings would then be comparable. The other possible solution is to design a beam-transmittance meter that does not have an inherent dependence on instrument geometry. This could be done by employing a dual-beam unit in such a manner that the comparison made would not be between the transmittance in air and water over the same pathlength, but between two similar water pathlengths, one greater than the other. If one pathlength were made twice the size of the other, the transmittance would reduce to

$$T = \frac{F_{L_2}}{F_{L_1}} \left(\frac{D + 2L_1 \tan \beta/2}{D + 2L_2 \tan \beta/2}\right)^2 = \frac{F_{L_2}}{F_{L_1}} \left(\frac{\frac{D}{L_1} + 2 \tan \beta/2}{\frac{D}{L_1} + 4 \tan \beta/2}\right)^2 \simeq \frac{F_{L_2}}{F_{L_1}} \left(\frac{1}{4}\right) \quad (12)$$

if $D/L_1 \ll 2 \tan \beta/2$ and $L_2 = 2L_1$.

In this dual-beam unit both beams would have the same diameter, source-spreading angle, receiver-acceptance angle, and receiver-aperture size (*see* Figure 7-15). Note that if this were done the result would be entirely free

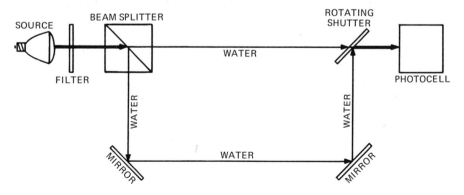

Figure 7-15. A possible dual-beam transmissometer. The rotating shutter would present either a neutral-density filter to direct beam or a mirror to reflected beam.

of any instrument-parameter effects; only the flux reaching the two receivers is included in the expression for the transmittance. This solution appears to hold more promise for the design of the future than does the standardization of all angular and length parameters of a single-beam unit.

From this development it may be seen that use of the laser solely for its small beam diameter and spreading angle does not have any particular advantage. However, the laser does have a distinct advantage in its almost discrete wavelength-emission property. For measuring beam transmittance at selected wavelength intervals, a tunable laser would probably be the ideal source. However, lacking a tunable laser of convenient size and power requirement, a prism beam-transmittance meter, or even a multiple-filter beam-transmittance meter, appears to be much more desirable than one that has no method of delineating source wavelength.

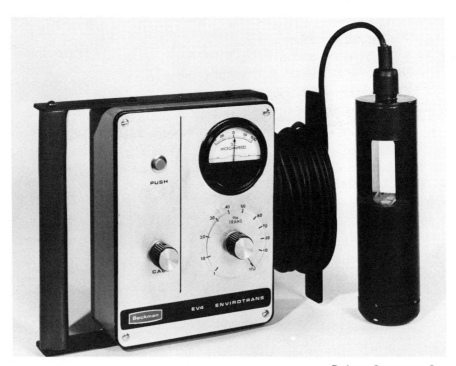

Beckman Instruments, Inc.
Figure 7-16. A portable beam-transmittance meter.

712. The Secchi Disc

One of the most common devices for determining water transparency is the Secchi disc. This is a white disc that is lowered into the water until it just disappears from view. The depth of disappearance is recorded as the Secchi disc depth, and this depth is supposedly related to the water transparency. Many attempts have been made to relate this reading to an absorption, scattering, extinction, or beam-attenuation coefficient, but none has been completely successful. This failure undoubtedly results from the complexity of the processes involved in seeing the Secchi disc.

A simple analysis of the disc, shown in Figure 7-17, shows that the light reaching the disc to illuminate it has been diminished by the process of extinction (relative irradiance loss). The light reflected from the disc to the eye, on the other hand, has been diminished by the process of beam attenuation because an image is being formed in the eye. Thus it may be seen that any relationship between the Secchi-disc reading and some loss coefficient of the water must include both the extinction coefficient and the beam-attenuation coefficient. One such relation (J. Williams, 1970) is

$$k + c = \frac{1}{D} ln \left[333 \left(\frac{R}{U} - 1 \right) \right] \qquad (13)$$

where k = extinction coefficient
c = attenuation coefficient
D = Secchi disc depth
R = disc reflectance
U = relative amount of upwelling light.

In addition to the two unknown optical coefficients in that expression, there is a third unknown, the relative amount of upwelling light. This unknown can be calculated if two Secchi discs are used, each having a different reflectance, R. White and black discs immediately suggest themselves for this purpose.

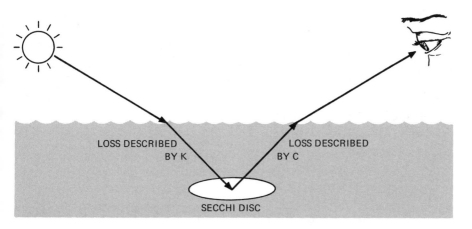

Figure 7-17. Different types of light loss experienced when a Secchi disc is used.

This is not the only improvement necessary in the use of the Secchi disc. The disc-eye-water-sun system is also an optical system with wavelength-response characteristics, just as is the light-water-photocell system in the beam-transmission meter. That being the case, it is necessary to specify the extinction and attenuation coefficients in terms of some particular wavelength, or at least, of some narrow wavelength band. It thus appears that in order to make Secchi-disc readings meaningful, somewhere along the line a filter must be placed in the system, and the easiest way of doing that is to have the viewer of the Secchi disc wear glasses containing a narrow-band color filter.

There is a great temptation to use the Secchi disc in preference to other optical instruments because it is so simple. However, the user must be constantly aware, even when he is using filter glasses and special viewing ports, of just what he is measuring. As indicated in Equation (13), the Secchi disc, at best, gives the sum of two loss coefficients which cannot be separated without the use of another type of optical instrument that measures either k or c uniquely.

713. Scattering Measurement

The direct measurement of scattering, as a function of both wavelength and scattering angle, has been tried by a number of investigators. However, none of the devices developed have found their way into the oceanographer's bag of routine measurement instruments. N. G. Jerlov (Jerlov, 1968) has designed, built, and used a number of nephelometers, all of which have essentially the same form. A typical nephelometer (*see* Figure 7-18) consists of a light source and a light sensor, such as a photocell, the sensor being fronted by a Gershun tube that can be aimed in many different directions. By rotating this Gershun tube-photocell unit around the center of the predetermined scattering volume, the variation of scattering with angle can be determined. In addition to this, filters can be placed on the light so that scattering can be determined as a function of wavelength also.

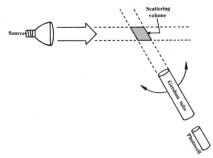

From *Optical Properties of the Sea*

Figure 7-18. Schematic representation of a typical nephelometer.

The basic principle that governs the design of a nephelometer is that a collimated beam illuminates a volume of water that scatters light in all directions. The volume of water that acts as a scatterer is determined by the volume common to the source beam and the imaginary beam produced by the receiver, as is shown in Figure 7-18. With this kind of instrument, secondary scattering effects make it difficult to determine the exact size of the scattering volume.

714. Measuring Bioluminescence

In recent years many other very sophisticated instruments for measuring light and light levels have been introduced, especially for scientific work with specific projects. One of these is a very sensitive device that measures bioluminescent flashes. This has been made possible by recent developments in multiplier phototubes and their associated solid-state amplifiers, which make it possible to measure light levels on the order of magnitude of 10^{-7} of that found at the water surface. This is about the level of threshold vision for the human eye and underwater television.

715. Integrating Devices

All of the aforementioned devices are used to describe an optical environment at a given time. However, when it is desirable to find the total amount of light reaching a particular depth of the ocean over a long period of time, integrated values can be produced. For example, photographic film, which in itself is an integrating device, has often been used: the optical density of film that has been exposed for a given period of time determines the total amount of incident light. Similarly, devices that are available commercially measure the total transported charge (coulombs) over a given period of time so that an integrated value of the total response of a photocell system at a particular depth can be obtained.

716. Light as a Tool

The use of light to measure oceanographic parameters from satellites has been suggested by many investigators. Mounting a laser aboard a satellite and using it for the accurate determination of water levels is a distinct possibility. In addition, ambient light can be used to determine water color and water clarity, especially in shallow areas where comparisons can be made.

Some work has been done in determining phytoplankton populations by the selective use of color photography. For that purpose, very-narrow-band filters are used in the hope of differentiating between the light reflected by the small plants and that reflected by the water itself.

Visual photography is also quite useful for determining ice cover and ice thickness. Lasers have been used to determine the relative height of different portions of the ice cover and, from this, ice thickness has been calculated.

Many of these methods have been utilized only from aircraft but there is no reason why they could not be extended to a satellite operation.

717. State of the Art

At the present time the state of hydrospheric optic instrumentation leaves much to be desired. No available instrument adequately measures beam transmittance, and it is difficult to determine what the present generation of transmissometers does measure. In contrast with the situation in many other disciplines, the most reliable devices in this field are those that are essentially the same as they were 30 years ago. It is time to redefine some of the basic parameters, and perhaps to do so in terms of what is presently measurable, rather than of basic theoretical relationships.

Sources and Additional Reading

Ackefors, H., G. Ahnstrom, and C. G. Rosen. "Construction and Performance of a Sensitive Light Meter for Underwater Use." *Limnol. Oceangr.,* 14(4), 1969.

Barrows, W. E. *Light, Photometry and Illuminating Engineering.* McGraw-Hill Book Co., New York, 1951.

Bauer, D., and A. Iyanoff. "Bathy-irradiance-mètre." *Cah. océanogr.,* 22(5), 1970.

———. "Spectro-irradiance-mètre." *Cah. océanogr.,* 22(5), 1970.

Brundza, R. "The Many Techniques in Using Camera Lenses for Underwater Photography." *Mar. Sci. Instrum.,* 4, 1968.

Clarke, G. L., G. C. Ewing, and C. J. Lorenzen. "Spectra of Backscattered Light from the Sea Obtained from Aircraft as a Measure of Chlorophyll Concentration." *Science,* 167(3921), 1970.

Duntley, S. Q. "Light in the Sea." *J. Opt. Soc. Amer.,* 53, 1963.

Hersey, J. B., ed. *Deep-Sea Photography.* The Johns Hopkins Press, Baltimore, 1968.

Hishida, K., S. Koizumi, and K. Nishiyama. On the Automatic Turbidity Measuring Instrument." (In Japanese: English abstract). *J. Oceanogr. Soc. Japan* 24(6), 1968.

Jerlov, N. G. *Optical Oceanography.* Elsevier Publishing Co., Amsterdam, 1968.

Karabashev, G. S. "Photometer for the Measurements of Irradiance Spectral Attenuation in the Sea." (In Russian; English abstract). *Okeanologiia, Akad. Nauk, SSSR,* 6(5), 1966.

Karabashev, G. S., and L. M. Nesterenko. "Field Logarithmic Photometer for Modulated Radiant Flux Recording." (In Russian; English abstract). *Okeanologiia, Akad. Nauk, SSSR,* 7(1), 1967.

Karelin, A. K., and V. N. Pelevin. "The Underwater Irradiance Meter FMPO-64 and Its Use for Optical Studies in the Sea." (In Russian; English abstract). *Okeanologiia,* 10(2), 1970.

Kaygorodov, M. N., and G. G. Neuymin. "Marine Polarimeter-Brightness Meter." (In Russian) In: *Metody i Pribory dlya Issledovaniya Fizicheskikh Protsessov v Okeane.* A. N. Paramonov, ed., Izd-vo Naukova Dumka, Kiev, 1966. Translation: JPRS: 39, 13 Feb. 1967. (Clearinghouse Fed. Sci. Tech. Info., U. S. Dept. Commerce.)

Klehn, H., and D. Sonntag. "Ein hydrographisches Extinktions- und Temperaturmessgerat." *Acta Hydrophys.,* 8(1), 1963.

Krause, G. "Eine Methode zur Messung optischer Eigenschaften des Meerwassers in grossen Meerestiefen." *Kiel. Meeresforsch.,* 19(2), 1963.

Middleton, W. E. K. *Vision Through the Atmosphere.* Univ. of Toronto Press, Toronto, 1952.

Neuymin, G. G., Ye. A. Agaforov, and S. V. Karaush. "Multipass Photometer-Transparency Meter." (In Russian). In: *Metody i Pribory dlya Issledovaniya Fizicheskikh Protsessov v Okeane.,* A. N. Paramonov, ed., Izd-vo Naukova Dumka, Kiev, 1966. Translation: JPRS: 39, 13 Feb. 1967. (Clearinghouse Fed. Sci. Tech. Info., U. S. Dept. Commerce.)

Paramonov, A. N. "A Marine Impulse Photo- and Transparency-Meter." (In Russian). *Okeanol., Akad. Nauk SSSR,* 4(2), 1964.

Patterson, R. B. "Photographic Techniques for Ocean Floor Search and Research." *Mar. Sci. Instrum.,* 4, 1968.

Rich, P. H., and R. G. Wetzel. "A Simple Sensitive Underwater Photometer." *Limnol. Oceanogr.,* 14(4), 1969.

Rosset, R. "Une camera de télévision qui 'voit' dans l'obscurité." *La Nature,* 1961.

Rusby, J. S. M. *Report on the Measurement of Refractive Index of Sea Water Samples.* UNESCO. Techn. Papers, Mar. Sci., No. 4: App. C., 1966.

Sasaki, T. "On the Instruments for Measuring Angular Distributions of Underwater Daylight Intensity." In *Physical Aspects of Light in the Sea,* John E. Tyler, Convener, Univ. Hawaii Press, 1964.

Sasaki, T., G. Oshiba, and M. Kishino. "A 4 π-Underwater Irradiance Meter." *J. Oceanogr. Soc.,* Japan, 22(4), 1966.

Secchi, P. A. "Esperimento por determinare la trasparenya del mare." *Cialqi Sul Moto Onaoso del Mare,* 1866.

Smith, R. C. "An Underwater Spectral Irradiance Collector." *J. Mar. Res.,* 27(3), 1969.

Stanton, L. W. "Comparison of Several Films for Underwater Use." *Mar. Sci. Instrum.*, 4, 1968.

Sugiura, Y., and Y. Akiyama. "Suitability of a Recording Photometer for Photometric Work Aboard the Observation Ship." *Oceanogr. Mag.*, Tokyo, 15(1), 1963.

Tyler, J. E. "In Situ Spectroscopy in Ocean and Lake Waters." *J. Opt. Soc. Amer.*, 55, 1965.

———. "The Secchi Disc." *Limnol. Oceanogr.* 13, 1968.

Tyler, J. E., and R. C. Smith. *Measurements of Spectral Irradiance.* Gordon and Breach Science Publishers, Inc., New York, 1970.

Westlake, D. F. "Some Problems in the Measurement of Radiation Underwater, a Review." *Photochemistry and Photobiology*, 4, 1965.

Williams, J. *Optical Properties of the Sea.* U. S. Naval Institute, Annapolis, 1970.

———. "Alpha Meter Design Considerations." *Trans. Instrum. Soc. Amer.*, 11, 1972.

CHAPTER 8

Sound Measurements

801. Sound Energy

As we have said previously, the ocean is a hostile, turbulent, heterogeneous medium. It is also an acoustic medium, since acoustic energy is the only form of energy that can be transmitted through an oceanic environment with acceptable losses. It is at least a thousand times easier to transmit sound energy through the ocean than to transmit electromagnetic energy through the ocean. For this reason, sound is used in the hydrospheric environment for many purposes, including search, identification, communication, data telemetry, and bottom and crustal investigation.

In attempting to design or use any sonic system, either for or in the oceanic environment, it is necessary to predict its operation. In order to do this we must know the effects of water on the passage of sound. These effects may be grouped as losses or noise, and we shall look at both.

802. Absorption

Losses generally fall into five catagories: those produced by absorption, scattering, spreading, refraction, and turbulence. Let us look at them individually and see how they can be measured in the final analysis. As in the case of light, we shall define absorption as the conversion of energy into heat, or the loss of sound energy by this method. Generally speaking, the absorption of sound in sea water is a function of the frequency of sound used. The dissolved magnesium sulfate in sea water appears to be the major contributor to this absorption loss. The absorption coefficient is related to the square of the frequency as given in

$$a = \frac{40f^2}{4100 + f^2} + 2.75 \times 10^{-4} f^2 \quad (1)$$

where a = absorption coefficient in decibels per kiloyard
f = sound frequency in kilohertz.

In this expression the decibel is defined as usual, that is

$$\text{sound level in decibels (db)} = 10 \log I/I_o \quad (2)$$

or $db = 20 \log p/p_o$ \quad (3)

where I = sound intensity
I_o = reference sound intensity
p = rms sound pressure
p_o = rms reference sound pressure (usually 1 dyne-cm^{-2} for work in water).

Equation (3) is true since, in general,

$$I \propto p^2$$

In Table 8-1 the absorption coefficient has been calculated for selected sound frequencies by the use of Equation (1). These frequencies are compared with the absorption coefficient in the same units for blue-green light, having a wavelength of 470 nanometers in air (350 nanometers in water). It may be seen that, especially for the lower frequencies, the absorption coefficient for sound is very much less than it is for light. Only when the frequencies get very large do the absorption coefficients approach the same order of magnitude as those for light.

Table 8-1 Comparison of Sound and Light Energy in Water

Energy type	Wavelength in water (meters)	Absorption Coefficient db/kiloyard
Blue-green light	3.5×10^{-7}	24.4
Low-frequency sound (400 Hz)	3.75	0.0016
Typical sonar (15 kHz)	0.1	2.14
Side-looking sonar (120 kHz)	0.0125	35.1

It might also be of interest at this point to compare the wavelengths of light and sound listed in Table 8-1. Even for a frequency of 120 kilohertz the wavelength of this sound in water is more than 100,000 times as great as the wavelength of blue-green light in the same medium. This, of course, is a built-in limit to the resolution possible for any sonic viewing system. Nevertheless, sound is used in an image-forming manner in both side-looking sonar and sonic holographic systems (*see* Section 807).

803. Scattering

Another type of loss experienced by both light and sound energy is scattering, or the redirection of the wave energy by scatterers in the path of the energy. Again, it can be seen why sound energy can be used for very dirty water. Reference to Table 8-1 indicates that the smallest wavelengths are on the order of magnitude of a centimeter or so, much larger than the particulate matter to be expected in suspension.

Particulate matter in suspension varies in diameter from about 0.5 μm up to about 0.5 mm, which is still well below the wavelength of sound used to penetrate dirty water. Most particulate matter is less than 0.1 mm; larger material tends to settle out quite rapidly. Scattering of sound energy is negligible compared to the scattering of light energy because light wavelengths are much closer in size to the material in suspension.

Sound energy, on the other hand, is scattered markedly by gas bubbles. It is apparently the resonance of the entrapped gas that causes these bubbles to scatter up to a thousand times as much energy as would be expected from a consideration of the simple cross-sectional area. Therefore, there is very often a large amount of scattered energy caused by bubbles within the surface layers, and this is an important aspect of what is generally called *reverberation*. Other sources of reverberation are:
1. convection cells, which cause juxtaposition of sinking cold water and rising warm water
2. wavy surface
3. turbulence accompanying waves that have gone beyond the limit of linearity (white caps)
4. variations in the characteristic impedance of the medium
5. micro-thermal structures
6. velocity microstructures
7. marine organisms
8. the sea bottom.

The largest sound return from unwanted targets usually comes from the surface layer as a reflection from the naviface (air-sea interface), entrapped bubbles, or from the ocean bottom. However, certain regions of the ocean contain a large biological community with sonic scattering properties called the *sonic scattering layer* (SSL). This scattering layer, comprised of organisms containing small gas bubbles, not only causes a great deal of unwanted sound scattering but also creates the illusion of a bottom where such does not exist. More than one well-defined SSL has been reported as a sea mount.

804. Spreading

Besides absorption and scattering, acousticians generally include in their loss calculations the concept of sound spreading. It is quite a bit more difficult to achieve collimation with acoustic energy than with light energy, so that in almost all cases spreading exists and contributes a non-negligible loss to the sound system. Of course, the energy is not lost; it is simply spread out over an area that increases with distance from the source. In this manner, the intensity, or the energy per unit area, is caused to decrease, and, since a hydrophone responds to intensity rather than to energy, the process is considered a loss.

The most common type of spreading is spherical spreading, wherein the energy spreads out as if it originated at a point source. As the distance from the source increases, the diameter of the sphere in which the energy is contained gets larger and larger. Since the surface area of a sphere is proportional to the square of its radius, spherical spreading results in a loss of intensity directly proportional to the square of the range. Expressed in decibels, spherical spreading is given by

$$\text{Spherical spreading (db)} = 20 \log R \tag{4}$$

where $R =$ the range.

If, however, the sonic projector is designed so that the energy spreads out as if the source were a cylinder, i.e., spreading occurs only in a horizontal direction, the resulting loss is termed *cylindrical spreading*. Since the surface area of a cylinder is proportional to its radius, cylindrical spreading produces a loss in intensity proportional to the first power of the range. Cylindrical spreading is given in decibels by

$$\text{Cylindrical spreading (db)} = 10 \log R \tag{5}$$

As will be seen later, spreading loss may be reduced by naturally occurring environmental conditions, as well as by transducer design.

805. Refraction

A loss of a somewhat different type is caused by refraction. Sound speed in the ocean not being a constant, sound generally travels in curved, rather than straight, lines. Consequently, an attempt to transmit sound energy from point "a" to point "b" in the ocean might fail completely because the sound energy was refracted, or bent away, from the straight-line path. Under such a condition, it would appear to an observer at point "b" that the losses were almost 100%. It would seem desirable, therefore, to have some knowledge of the variation of sound speed within the oceanic environment if the probability of getting a reasonable amount of sound energy to a particular target area is to be accurately predicted.

The speed with which sound energy travels through the oceanic volume averages about 1,450 meters per second at the surface and increases as either salinity, temperature, or pressure increases. The best empirical relationship we have at the present time for describing this variation is Wilson's and it is given in

$$c = 1449.14 + V(p) + V(S) + V(T) + V(STp) \text{ (meter/sec)} \tag{6}$$

The variation in sound speed due to pressure different from atmospheric is $V(p)$, that due to salinity different from $35°/_{oo}$ is $V(S)$, that due to temperature different from 0°C is $V(T)$, while $V(STp)$ is the variation produced by the combination of all three parameters. By tabulating these four anomalies, sound speed can be determined with a great deal of accuracy, if precise temperature, salinity, and depth are known.

The amount of speed change produced by an incremental change in the three parameters just mentioned varies a great deal. A temperature increase of 1°C causes sound speed to increase by between 3 and 4 meters per second, depending upon salinity and pressure values. A salinity increase of 1°/$_{oo}$, on the other hand, increases sound speed by only about 1 meter per second, while a depth increase of 55 meters brings about an increase in sound speed of 1 meter per second. In the deep ocean, maximum spatial variations on the order of 3°/$_{oo}$ to 4°/$_{oo}$ in salinity may be occasionally experienced, while a temperature difference of as much as 25°C may be occasionally found between surface and bottom waters. Thus, temperature differences alone can cause sound speed to vary with depth by as much as 100 meters/second. Salinity differences can be expected to engender sound speed variations almost two orders of magnitude smaller. It is obvious, then, that temperature-induced changes in sound speed are so much greater than those generated by salinity differences that salinity effects may often be neglected. In predicting sonar ranges, for example, only temperature and pressure (depth) effects are considered.

Thus, the speed with which sound energy travels through the hydrospheric environment is seen to be related to the parameters of the environment. It should be emphasized at this point that these are the only parameters to which sound speed is related. Sound speed does not vary with frequency; a measurement at any one frequency suffices for all.

Since temperature and depth are the major determiners of sound speed in the deep ocean, sound speed variation may be predicted from a determination of these two parameters. Common variations of temperature with depth are those in which the upper layer of the ocean is either isothermal or else decreases slightly as the depth increases. At a somewhat lower layer, as has been shown previously, there is usually a marked decrease of the temperature with depth, called a *thermocline,* and below this another layer which is relatively isothermal.

Given those conditions, sound speed can be expected to vary with depth, as indicated in Figure 8-1. In the upper isothermal layer, increasing pressure causes an increase of sound speed with depth; while in the thermocline, decreasing temperature overrides the effect of increasing pressure and generates a decrease of sound speed. In the deeper layer, where the temperature gradient again becomes small, increasing pressure will once more occasion an increase of sound speed. Under these conditions, then, a sound speed maximum is found at the beginning of the thermocline, and sound traveling above the thermocline tends to be bent upwards. However, there will be some sound rays emitted from a near-surface source that will not be bent upwards enough and will penetrate into the region of decreasing sound speed. The most horizontal of these penetrating rays is called the *critical ray*, and between it and the lowest upward-bending ray there is a region of very little sound energy penetration, called the *shadow zone*.

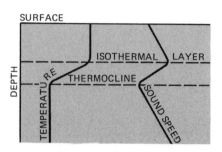

Figure 8-1. Variation of sound speed with depth for a mixed-layer above a thermocline.

Above the shadow zone, sound energy is confined within a duct due to refraction at the base of the duct and reflection from the water surface on the top. As is shown in Figure 8-2, spherical spreading is reduced to cylindrical spreading and ranges are quite great within the surface duct.

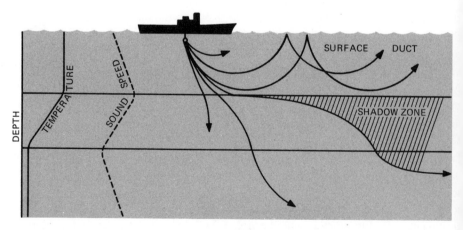

Figure 8-2. Formation of a surface duct and a shadow zone.

With a further increase in depth, a sound speed minimum is reached at approximately the bottom of the thermocline. At this sound speed minimum there is the sound-channel axis, from which sound traveling upward tends to bend downward and sound moving downward tends to bend upward. Sound energy is confined within two horizontal planes, as may be seen from Figure

8-3. In this case, confinement at both top and bottom is caused by refraction, so that within this sound channel ranges can be exceptionally great if the source is placed at the channel axis. These examples show that environmental conditions can reduce spreading losses, while, at the same time, refraction can cause extremely large losses.

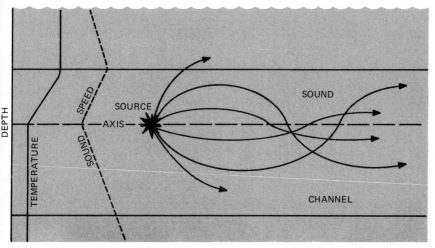

Figure 8-3. Formation of a sound channel.

806. Time-Dependent Losses

Finally, there is sound-energy loss associated with temporal changes within the environment itself. Such changes may be very short or relatively long in duration. They may be the result of small parcels of turbulence evoking reverberation, or they may be tied to internal waves that cause the interface between the mixed layer and the deep layer to change depth with a periodicity on the order of 10 minutes or more. They may also be caused by the movement of particulate matter or bubbles in sea water which do not necessarily stay in one position. This is especially true of particulate matter that is alive (zooplankton, for instance) and of bubbles associated with members of the nektonic community (fish, for instance). An excellent example of this is the sonic scattering layer found in the Atlantic Ocean, where the major source of sound scattering is the swim bladders of small fish. Fish obviously swim continually and provoke a reverberation that changes with time. The effect created is especially annoying when an attempt is being made to do imagery by means of sonic energy.

807. Acoustic Holography

One way of getting around this problem of variability might be holography, a three-dimensional imaging system that uses a portion of the ensonifying energy as a reference with which to compare the sound scattered back from the object. The diffraction pattern resulting from these two beams is recorded as a holographic image. There are two major advantages to this method.

1. The information appearing on the holographic print is three-dimensional in character and, when illuminated in the proper manner, it presents a three-dimensional picture, so that the observer has the illusion of looking at the real scene through a window. If he moves about, he can even look behind objects in the hologram.

2. Each small portion of the image presented in the hologram contains sound energy from the entire object, so that the effects of any individual particles tending to cause reverberation are smeared over the entire image. In this manner, the effects of scattering can be minimized.

There are some basic problems in developing sonic holography, one of them being the wavelength of the energy utilized and the conversion of this to optical wavelengths so that they can be seen with the eye. The wavelengths are so different that, theoretically, there should be a demagnification in the same ratio as that of the wavelengths of sound to light, in order to produce the same amount of resolution as would have been produced had light been used to begin with. Since it does not seem practical to make that adjustment, poor resolutions would have to be accepted. Nevertheless, since these poor resolutions may be adequate for some purposes, it is still necessary to know some of the sonic properties of the water before holographic pictures can be made.

808. Ambient Noise

In the utilization of sound energy within the oceanic environment, the problem of ambient background noise is ever present. Such noise, whose level can be high, is generated by thermal effects, surface activity, human activity, and biological sources. In deep water, where the only important sources of noise are thermal and surface, ambient noise levels have been calculated by Martin Knudsen and presented as a set of curves (*see* Figure 8-4). Note that Knudsen's curves describe the spectrum level in decibels as a function of the frequency of sound. Note also that Knudsen's curves go up only to sea state 6, which is described as a high sea with waves between 12 and 20 feet. Such a sea would probably be produced by a wind speed of about 30 knots. There are, of course, stronger winds blowing at sea, such as are experienced in a gale or a tropical storm (hurricane or typhoon), but no data have been taken on them, so the best that can be done is to extrapolate to greater sea states. Nevertheless, the data that have been taken show a steady decrease in sound energy with frequency increase for these naturally produced noises.

In a harbor or a coastal area where shipping is present, sound energy rather markedly depends on frequency. Figure 8-5 shows three typical curves: one for heavy shipping, one for moderate shipping, and one for light shipping. All three show a tendency toward greater sound levels in lower frequencies compared to those associated with the wind.

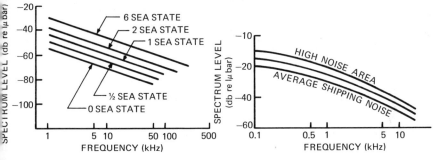

Figure 8-4. Ambient noise for sea states from 0 to 6.

Figure 8-5. Shipping-noise spectra.

Lastly, Figure 8-6 shows data for two common biological sound-producers, croakers and snapping shrimp. Croakers are fish that make bullfrog-type sounds, and they are common to coastal areas. Snapping shrimp are small shrimp that congregate in large numbers and, by rubbing claws together, produce a snapping noise similar to that made by an individual snapping his fingers. Note that the spectrum produced by snapping shrimp is relatively flat right across the frequency band, while that produced by croakers seems to be peaked pretty much in the same frequency band as shipping and sea-state noises. In any event, it is possible for one of these sources alone to produce a background noise level that makes the signal-to-noise ratio impossibly small for any useful work.

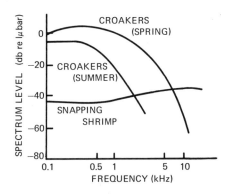

Figure 8-6. Typical fish-noise spectra.

809. Geophysical Sound Sources

Devices that can either produce or receive sound underwater are required for measuring sound losses and ambient noise, and the type used depends upon the sound desired.

For geophysical exploration, dynamite, gas exploders, and mechanical, pneumatic, and electrical contrivances have been used.

Dynamite has the major advantages of being relatively inexpensive and of producing a rather large amount of acoustic energy that falls within a relatively small band of low frequencies generally peaked within the region of least absorption by the sea. Lately, however, there has been a tendency to steer clear of dynamite on account of its being potentially dangerous.

Gas exploders use propane and oxygen to effect a detonation.

Mechanical devices generally rely upon the motion of a metal plate that generates sound waves, much as a diaphragm does in a loud speaker. One such device is the thumper, which consists of an electric coil that generates eddy currents in a metal plate when a capacitor is discharged through the coil. These eddy currents evoke a repulsive force caused by the opposing induced magnetic field, and the plate is rapidly moved away from the coil. A thumper can be used over and over again by simply recharging the capacitor and refiring.

Pneumatic devices in use today usually include a chamber into which air is forced at high pressure, a shutter that keeps the air contained, and a solenoid-actuated valve that gives high-pressure air access to the shutter in such a way that it disturbs the shutter's equilibrium and trips it. Once the shutter has been tripped, air rushes into the water with explosive force and gives rise to an acoustic signal. Different-sized units can be used in a series so that a number of different frequencies can be generated at the same time.

Electrical devices obtain large amounts of acoustic energy in the low frequency band by producing a large-voltage spark underwater. Sparkers can be fired at frequent intervals, but they require a fair amount of electrode maintenance.

For data transmission, communication, and object search and identification, the most commonly used sources are electronic ones and these will be discussed in Section 810.

810. Acoustic Transducers

All of the aforementioned devices produce a pulse of energy that contains a relatively narrow band of frequencies and is sent into the medium and received either close by or far away from the transmitter. A device intended for studying the sonic properties of the ocean medium must have more control on the source. For this reason, transducers capable of generating sinusoidal sonic energy within a predictable beam pattern have been developed. Three types of materials, classified as magnetostrictive, piezoelectric, and electrostrictive, have been used in sonic transducers.

Magnetostrictive means either that the physical dimension of a material is changed when the material is influenced by a magnetic field, or a magnetic field is produced when the material undergoes physical stress. A transducer that uses this property has a coil wound about a magnetostrictive alloy rod attached to a plate. When current is passed through the coil, the resulting magnetic field causes the plate to vibrate at a rate determined by the frequency of the electrical current. In the receiving mode, when the plate moves, the force that is exerted on the rod induces a current in the coil. This phenomenon is exhibited by the ferromagnetic metals, iron, nickel, and cobalt, and many of their alloys. Although magnetostrictive transducers are not as popular as they once were, they are still used for high-power, low-frequency projectors.

Piezoelectric is the term used to describe the effect created when the stressing of certain crystals, such as quartz, produces an electric polarization proportional to the magnitude and direction of the strain.

Electrostrictive describes the process whereby a dielectric, when subjected to an electric field, undergoes a physical deformation, commonly called electrostriction, which is independent of the direction of the electric field and is proportional to the square of the induction. It does not, therefore, change sign with the exciting electric field.

The electrostrictive effect is distinguished from the piezoelectric effect in two respects: it is common to all materials; and it is usually much smaller than the piezoelectric effect. In fact, it is so much smaller that ordinarily it can be ignored even though it is always present when the piezoelectric effect is manifested. A notable exception occurs with the ceramic barium titanate, for which the electrostrictive effect is large in comparison to the piezoelectric effect. In recent years other ceramics with extremely large electrostrictive effects have been developed and they are tending to replace piezoelectric materials in many applications.

Table 8-2, which presents the properties of the more common piezoelectric and electrostrictive materials, shows that the new ceramics are not better in every respect than the older crystals. For example, ADP still has the greatest voltage per unit stress value of any of the materials listed, while PZT has the highest maximum working pressure and the highest maximum electric energy output of any of the materials listed. Thus, once more, a designer must determine under what environmental conditions his device will be operating, and that determination plus the size, power, and other considerations of the device enable him to choose the proper materials.

One of the advantages of both piezoelectric and electrostrictive materials is that small and specially shaped transducers can be rather easily fabricated.

Table 8-2 Properties of piezoelectric and electrostrictive materials

Material	Energy Conversion (%)	Dielectric Constant	Young's Modulus 10^6 psi	Voltage per unit stress (q)	Max. working pressure (psi)	Max. Electric Energy Output (joule/in^3)
Piezoelectric materials						
Rochelle salt	80	350	3	90	1,000	0.01
Ammonium-dihydrogen-phosphate (ADP)	10	15	3	180	3,000	0.015
Quartz	1	4	12	58	50,000	0.12
Electrostrictive materials						
Barium titanate	15–25	500–2,000	17	10–15	5,000	0.02
Lead titanate zirconate	25	600	12.5	24	10,000	0.12
PZT*	35–50	350–2,500	8.5–14	20–40	to 20,000	to 0.6
Lead niobate	18	250	5	40	5,000	0.03
Desirable	High	Depends on load impedance	Low	High	High	High

* A type of lead titanate zirconate.

It is possible, therefore, to develop small projectors with narrow beam patterns for higher-frequency sonic applications, because the directionality is directly related to the ratio between sound wavelength and projector diameter. The smaller the wavelength with respect to the size of the projector, the more directional is the beam. Of course, as can be seen from Table 8-2, when large power outputs are required, as is often the case in the newer operational sonar gears, large amounts of crystalline or ceramic material have to be used. The largest value of energy output per unit volume is exhibited by PZT, and that is only a little over half a joule per cubic inch. When it is considered that modern sonars have energy requirements on the order of magnitude of kilowatts (1 kilowatt = 1,000 joules per second), it can be seen why transducers must be large.

811. Transducer Calibration

Almost any experiment involving the propagation of underwater acoustic signals requires a knowledge of the propagation loss a signal experiences as it travels from the point of its origin to the point where it is received. Such knowledge is a basic requirement for the designer of equipment that uses underwater acoustic signals for such purposes as depth determination, fish finding, and underwater acoustic communication. Underwater acoustic sources that generate sinusoidal signals are the conventional means of measuring propagation loss in the ocean. The amplitude of the propagated signal is then measured as a function of distance from the source, and usually determined in selected frequency bands by squaring and integrating the output of a filter fed by this extremely broad transient signal.

Because of its very hostile environment, the underwater sound transducer must meet stringent requirements. It has to be capable of producing or detecting sound pressures in an environment that may impress a hydrostatic pressure one billion times the signal amplitude and still not have its performance significantly degraded. Measurements are generally taken either in the far field of a sound source, that is, the region where the sound waves are spherically divergent, or at a distance greater than the ratio of the source aperture to the sound wavelength.

When the pulse technique is used, the output of the hydrophone is measured while the acoustic pulse arriving from the straight-line path is impinging on the hydrophone. These tests are usually conducted in a tank, where the reflection of sound from the sides, top, and bottom of the tank interferes. However, since the dimensions of the tank are known, most of the reflection can be eliminated by electronically gating the receiver and allowing only direct-path reception. This is often done by placing the source and the hydrophone at half the depth of the water in the tank.

It should be emphasized that, if the operating environment is to be properly simulated in a test tank, the water should be pressurized and temperature-controlled. If a large-volume pressurized tank is not available, the required conditions can be simulated by enclosing the transducers in acoustically transparent vessels suspended in an open test tank. Fiberglass, other plastics, and high-pressure hose have been used for this purpose. Non-acoustic hydrostatic tests are not good enough for testing transducers because, while they may reveal a structural weakness, they may fail to reveal performance degradation.

Near-field measuring techniques that reduce to a few inches the distance required between the source and the hydrophone have been developed. These techniques extend materially the capabilities of existing test facilities for calibrating large transducers. A planar array of small sources can produce a plane wave front of constant amplitude over the finite volume that contains the measuring transducer. The same array can integrate the near-field radiation from the measured transducer and yield data that show the far-field radiation characteristics. A single small transducer or a line array can be used to obtain the same data from its components.

The following are a few of the many problems that arise when piezoelectric· or electrostrictive transducers are used in the deep-ocean environment:
1. Lowered electrical leakage resistance caused by the penetration of water vapor
2. Changes of frequency response and directivity as depth changes
3. Acoustic properties as a function of operating depth change before the ultimate strength of the materials is exceeded
4. Acoustic properties are apt to exhibit the effects of hysteresis as hydrostatic pressure is cycled.

For these reasons, acoustic tests under actual or simulated environmental conditions are vital in the development of underwater sound transducers.

812. Propagation Loss

For measuring sonic environmental characteristics with acoustic transducers, the receiving transducer is placed at some distance from a calibrated source. If spherical spreading is assumed, the sound-pressure level in decibels at the source if then given by

$$L = 20 \log P_o - 20 \log R - aR - TA \tag{7}$$

where P_o = sound pressure at one yard from the source
R = distance from the source in yards
a = absorption coefficient in db per yard
TA = transmission anomaly, including the effects of refraction and reverberation.

Since this sound-pressure level is also given by

$$L = 20 \log p \tag{8}$$

where p = sound pressure at the receiver hydrophone

and $p = V/S$ \hfill (9)

where V = voltage generated at the hydrophone terminals
S = hydrophone sensitivity in volts/dyne/cm^2

then $20 \log p = 20 \log V - 20 \log S$ \hfill (10)

so that $20 \log V = 20 \log P_o - 20 \log R - aR + 20 \log S - TA$ \hfill (11)

Thus, the sensitivity of the receiver being known, the characteristics of the environment can be determined by measuring the output of the source.

813. Self Noise

Another difficulty that enters into any acoustic measurement at sea is the presence of self noise. Self noise is the noise produced by a vessel and any of its accessory equipment, as measured by a signal-detecting, tracking, or measuring hydrophone on the vessel. Self noise includes any noise inherent in the detecting, tracking, or measuring system, the level of which is high enough to be significant. However, when the vessel carrying the equipment is being propelled through the water, system noise can normally be reduced to an insignificant level. Consequently, most self noise is either waterborne or structure-borne noise that enters the system through the hydrophone. Other noise may be generated in the electronic equipment itself or be ambient noise such as has been discussed above. All of these must be considered in connection with any measurement taken at sea.

814. Sound-Speed Measurement

In order to assess the effects of refraction on sound losses, the variation of sound speed within the ocean has to be determined, and there are two ways in which this can be done. One is to measure the sound speed with a sound velocimeter; the other is to measure the salinity, temperature, and depth and, from those results, calculate the sound speed by Wilson's equation [Equation (6)]. At this time, the latter is probably the more accurate of the two methods, but since the former is more convenient we shall discuss sound velocimeters in some detail.

Sound speed can be measured to an accuracy adequate for operational use at sea. This is done characteristically by the use of the sing-around circuit developed by Greenspan and Tschiegg at the Bureau of Standards in 1957. Their system consists of two piezoelectric or electrostrictive transducers separated by a known distance and forming a fixed water-sound path. A blocking oscillator generates a pulse, allowing the transmitting transducer to vibrate at its resonant frequency for the duration of the pulse. This sound pulse travels over the fixed path to the receiving transducer where it is converted into an electrical impulse that then re-triggers the blocking oscillator, starting the whole sequence again. The time between pulses, therefore, is a function of the time of transit of the sound energy and any electronic delays in the circuit. Since electronic delays are usually very small, sound speed is determined by the pulse-repetition rate of the system.

One of the inherent problems with this type of system is that it is difficult to determine the point at which the blocking oscillator is re-triggered. The leading edge of the pulse must somehow or other be sampled, so that the re-triggering occurs at the same portion of the pulse all the time. Unfortunately, the attenuation characteristics of the sample have a direct bearing on the shape of the pulse and, therefore, on the location of the triggering point. As was seen before, the absorption of sound energy is directly related to its frequency. An increase of frequency produces increased absorption. It follows, then, that this selective attenuation alters the shape of a pulse front that contains energy over a wide high-frequency range.

Furthermore, the absorption of sound energy is inversely related to the temperature: the lower the temperature, the greater the amount of absorption it produces. At 15 megahertz, attenuation for a 20-centimeter path is almost 10 decibels greater at $2°C$ than at $30°C$, while at 30 megahertz the difference in attenuation between these temperature limits is more than 30 decibels. Thus it can be seen that changes in water temperature introduce significant errors in the system. In order to compensate for this, the received pulse is usually amplified and re-shaped before it triggers the blocking oscillator.

Sound-speed measurement is, of course, affected by the speed of water through the sound path. To minimize this effect, the sound path is folded in either the horizontal or the vertical plane by an intermediate teflon, hard-rubber, or perforated-metal reflector. Likewise, since the sound signal can be reflected from the receiver back to the transmitter and back to the receiver, several methods are used to eliminate the effects of multiple echoes between transducers. One of these methods is to tilt the transducers out of parallel.

Transducers and reflectors must be mounted on plates that ensure the maintenance of a pathlength that is relatively constant with temperature change. Invar is commonly used to minimize the effects of temperature on the sound pathlength, but it is subject to corrosion. Another approach is the selection of metals with dissimilar temperature coefficients while a third method permits computation of pathlength by knowledge of ambient temperature and the coefficient of expansion of the plates.

Air bubbles on transducer faces, which often occur when temperature changes rapidly, cause high attenuation of sound pulses and, consequently, erratic operation of velocimeters. Also, long immersion at high pressures has been known to cause water seepage behind the transducers, resulting in velocimeter failures.

Amplifier bandwidth and gain are also significant factors in the operation of the sing-around system. Amplifier bandwidth is dictated by the rise time of the pulse, the acoustic pathlength, and the accuracy requirement; a wide bandwidth is required for fast rise time, short pathlength, and precise measurements. Electronic time delay and calibration non-linearity can be reduced by increasing amplifier bandwidth.

At the present time, devices for measuring sound speed appear to have an over-all routine accuracy of about ±0.4 meters/second. That accuracy is limited by considerations of temperature compensation, pressure effects, fouling, and corrosion.

815. State of the Art

Thus, we see that in all cases the measurement of sonic properties of the ocean is very much the same as that of optical properties. The measurements are relatively crude, but more reliable and more accurate instruments will become more readily available in the very near future.

Sources and Additional Reading

Beccasio, A. J. "Dye Markers for Sonobuoys." *Undersea Technol.,* 5(8), 1964.

Buchanan, C. L. "Wide-band Transducers for Sound Velocimeters." *Mar. Sci. Instrum.,* 2, 1963.

Carnvale, A., et al. "Absolute Sound-Velocity Measurements in Distilled Water." *J. Acoust. Soc. Am.,* 44(4), 1968.

Francis, T. J. G. "A Long-Range Seismic-Recording Buoy." *Deep-Sea Res.,* 11(3), 1964.

Greenspan, M., and C. E. Tschiegg. "Sing-Around Ultrasonic Velocimeter for Liquids." *Rev. Sci. Inst.,* 1957.

Haslett, R. W. G., G. Pearce, A. W. Welsh, and K. Hussey. "The Underwater Acoustic Camera." *Acustica,* 17(4), 1966.

Kroebel, W., and J. Wick. "Registrierungen in situ im Nordatlantik mit der Bathysonde und einem neuen Messgerät für Schallgeschwindigkeit im Meerwasser mit extrem hober Genauigkeit." *Kiel. Meeresforsch.,* 19(2), 1963.

McSkimin, H. J. "Velocity of Sound in Distilled Water for the Temperature Range 20°–75°." *J. Acoust. Soc. Am.,* 37, 1965.

Morand, C., M. Teton, and G. Lejeune. "Description et construction d'une sonde thermoélectrique pour la measure de l'intensité acoustique dans l'eau." *C. R. Acad. Sci.,* Paris, 257(5), 1963.

Trott, W. J. "Underwater-Sound-Transducer Calibration from Near-Field Data." *J. Acoust. Soc. Amer.,* 36(8), 1964.

U. S. Naval Oceanographic Office. *Tables of Sound Speed in Sea Water.* Spec. Pub. SP-58, Washington, D. C., 1962.

Urick, R. J. *Principles of Underwater Sound for Engineers.* McGraw-Hill Book Co., New York, 1967.

Van Reenan, E. D. "A Complete Sonar Thumper Seismic System." *Mar. Sci. Instrum.,* 1, 1962.

Weber, H. P. "An Experimental Near Real Time Sound Velocity Profile System." *Mar. Sci. Instrum.,* 4, 1968.

Wilson, W. D. "Equation for the Speed of Sound in Sea Water." *J. Acoust. Soc., Am.* 32, 1960.

CHAPTER 9

Chemical Measurements

901. Use of Chemical Measurements

For many years, the chemical measurements made at sea have been used as much by physical as by chemical oceanographers. As examples, dissolved oxygen data have been used in efforts to determine the gross circulation of the oceans; and it has been routine to use chlorinity determinations to calculate density, from which the current structure of the oceans could be determined. In recent years, however, more and more of the physical parameters have been measured directly. Oxygen measurements for calculating ocean circulation have been largely replaced by the use of other more conservative properties of water masses, such as radioactive carbon, and density is now determined from a conductivity measurement.

Chemical oceanography has finally reached the point where the work involved is important for its own sake. Phosphates, nitrates, silicates, and many other dissolved chemical constituents at sea are now measured as a matter of routine, the most commonly measured being oxygen and pH. In this chapter we shall discuss the measurement of these two parameters, since *in situ* electrodes are more readily available for their determination.

902. Dissolved Oxygen Distribution

Dissolved oxygen exists throughout the oceanic water column. It varies in concentration, the maximum being 9 milliliters per liter, equivalent to 13 parts per million by weight. Since, generally speaking, the ocean derives all its oxygen from the atmosphere and the end product of photosynthesis, all of the oxygen is put into the ocean in the upper layers. Therefore, if water at depths below about 100 meters contains any dissolved oxygen, that water must, at some time, have been at or near the surface.

Oxygen in the ocean is depleted by the respiration of animals and by biological decay. These two processes are continually at work, so that unless a source of oxygen were continually available, the oxygen supply of the ocean would decrease to zero in a short period of time. Apparently, however, there is an equilibrium between the amount of oxygen supplied by the plants and the atmosphere and the amount consumed by respiration and biological decay.

The maximum amount of dissolved oxygen that can be maintained in solution under saturation conditions is a function of salinity, temperature, and pressure. Saturation depends on temperature and salinity in such a way that cold fresh water can hold more oxygen in solution than can warm saline water. With a salinity of $0°/_{oo}$ and a temperature of $0°C$, water can hold about 14 parts per million of oxygen, whereas with a salinity of $30°/_{oo}$ and a temperature of $30°C$, it can hold only about 6.3 parts per million. Thus, changes in salinity and temperature markedly affect the amount of oxygen that can be held in the water and, therefore, near the surface, the amount of oxygen actually present.

Upper layers of the ocean are often supersaturated, and values as high as 130% of those indicated above are not unusual. This is due to the strong mixing processes whereby saturated water is carried to greater depths with increased pressure, so that at these depths, with a given temperature and salinity, the saturation condition is exceeded.

Since both oxygen sources are in the surface regions, it would be expected that most of the oxygen present in the ocean would be in this region, and this is generally true. Figure 9-1, a typical oceanic plot of oxygen *vs* depth, shows a relatively high value of oxygen at the upper layer and, just below the surface, where the supersaturation phenomenon occurs, a still higher value. When the compensation depth is reached (the depth at which the oxygen input is equal to its consumption), the amount of dissolved oxygen decreases continuously to approximately the depth of the thermocline, where it is very common to find an oxygen minimum. Between the thermocline depth and some intermediate depth, the oxygen level increases again before decreasing near the bottom.

The reason for this oxygen minimum is apparently that very little vertical mixing occurs at thermocline depth: if there were more, of course, the thermocline would be erased. Below the thermocline, then, oxygenated water that has sunk from the surface is carried in by horizontal motion and increases the value of oxygen present. There not being much food at these depths, the amount of animal life is not quite as great as it is in the surface layers. Consequently, the demands on the oxygen supply are somewhat less than they are in the surface layers.

Oxygen is present in all water layers, even at the greatest depths. In 1960 when the *Trieste* dived to the bottom of the Mariana Trench, at a depth of 35,800 feet she found measurable amounts of oxygen.

Conditions in some estuarine, coastal, and lake areas are in direct contrast to those of the open ocean. In the former regions, where the influence of man is felt very strongly, there are periods when oxygen supplies are greatly diminished by the amount of biological decay caused by waste products in the environment. In certain areas, especially during the summer when the water is warm and decay processes are accelerated, there may be little or no oxygen at depths that are not too great.

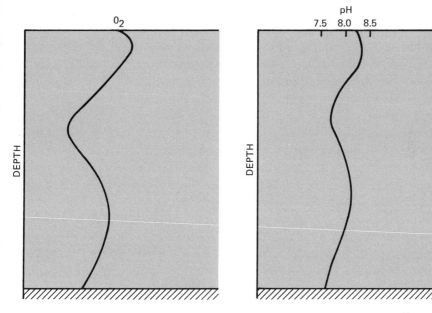

Figure 9-1. Idealized plot of oceanic dissolved oxygen vs depth.

Figure 9-2. Typical plot of oceanic pH vs depth.

Furthermore, natural conditions, such as those that exist in the summer months when the winds are very light and there is little or no mixing, often create a similar situation. With large amounts of animal life using the oxygen, there is no mechanism for replenishment at depth. Thus, in certain areas, it is common during July and August to find values of oxygen less than 1 part per million at depths below 25 or 30 feet.

903. Oxygen Electrodes

The classical way of measuring dissolved oxygen is by the Winkler chemical titration, the standard against which all other devices or processes are compared. However, a number of the electrodes that have been introduced in recent years hold promise of being able to measure dissolved oxygen values *in situ*. Among these are the polarographic and the thallium-metal electrodes.

A polarographic electrode consists of a gold cathode and a silver anode surrounded by a gel electrolyte. A small voltage applied between the electrodes makes the sensor specific to oxygen. In the majority of designs, a permeable membrane (usually a thin sheet of teflon or polyethelene) covers the sensor element in such a way that oxygen diffused through the membrane is reduced electrochemically at the cathode, producing an electrical current proportional to the dissolved oxygen concentration.

In one design, however, the sensor uses a relatively impermeable diffusion membrane which limits the electrode's oxygen consumption to a rate near that of local diffusion in the sample. The result is that the sensor is relatively insensitive to the variations in water velocity near the electrode. This problem is inherent in this type of device because the oxygen measured is actually consumed in the electrochemical process of measuring it. Unless the fluid being measured is constantly replaced, the measured value of dissolved oxygen continues to decrease as the oxygen in the fluid is rapidly decimated in the region of the electrode. Unfortunately, if the sensor is designed to measure a very small portion of the dissolved oxygen, its time constant is increased.

A thallium-metal electrode operates in a similar manner, but a slightly different reaction occurs in that when thallium is exposed to oxygen, a concentration of thallous ion proportional to the oxygen concentration is produced at the surface of the electrode.

Polarographic electrodes are the ones most commonly used to measure oxygen. Since this type of electrode requires a constant diffusion of oxygen through the permeable membrane, it is imperative that the membrane be kept clean. Fouling of any kind is unacceptable because it would make the device inoperative. For this reason, buoys are not the best platforms for these units, although there have been schemes suggested including wipers activated periodically to keep the membrane clean.

As indicated previously, the time constant of this type of electrode is essentially a function of the rate oxygen is diffused through the membrane. If the primary interest is time constant, then the membrane must be as thin and as porous to oxygen as possible. An electrode with a 0.001-inch-thick polyethelene layer will show a 90% response in about 15 seconds, while a 0.00025-inch teflon membrane will show a 90% response in 3 seconds. However, the thinner the membrane, the more fragile and, of course, the greater the diffusion rate of oxygen through the membrane the greater the mechanical mixing required to maintain a fresh supply of oxygen for measurement.

At this time, accuracies of the polarographic electrode seem to be about ± 0.03 milliliters per liter when the devices are kept moderately clean. However, sensors used on buoys apparently show some degradation in accuracy caused by fouling.

904. pH Distribution

Sea water is normally alkaline. In the surface layers, it has a pH that varies between about 8.1 and 8.3, and in the lower layers the value decreases to somewhere around 7.5. However, in diluted water and isolated basins where hydrogen sulphide is produced, the pH may approach 7 or even fall in the acid range. This is particularly true in estuarine and coastal regions where there are large quantities of biological decay and both domestic and industrial waste products.

In the open ocean, the pH is essentially a function of the amount of dissolved carbon dioxide present and it decreases as the amount of CO_2 increases. Thus, in regions of high photosynthesis where there are large amounts of oxygen, the pH is relatively high; while in regions where there is a lot of animal respiration and biological decay, the pH is relatively low. Usually, therefore, oceanic variation of pH with depth is similar in a qualitative sense to the variation of oxygen with depth (*see* Figure 9-2). The pH is normally greatest in areas of high oxygen concentration and smallest in areas of low oxygen concentration.

Either an indicator or an electrode is usually used for the measurement of pH. In both cases an unknown is compared with a standard in order to determine the pH.

905. Indicator Methods for pH Determination

Indicator methods for the measurement of pH utilize color variations related to the pH. An unknown solution can be compared with known solutions visually or spectrophotometrically, depending on the accuracy desired. Thus, the pH of an unknown solution is the same as a buffer solution with the same color. In any pH measurement method temperature correction or compensation is required if the temperature of the sample is different from that of the reference, because the pH varies with temperature.

906. Electrode Methods for pH Determination

The electrometric method uses electrode potentials to find the unknown pH by comparing the measuring electrode with a reference electrode. Electrodes that have been utilized are hydrogen, quinhydrone, and glass.

Hydrogen electrodes need to have carbon dioxide continually flushed from the sample, because a stream of hydrogen is required for proper operation. Unless this is done, the reduction of carbon dioxide in the sample solution causes the pH to increase. However, when proper care was taken hydrogen electrodes have been used successfully as a laboratory standard.

Quinhydrone electrodes can be used only when the pH is less than 8.5, a minor limitation for most oceanic applications. At low pH values, however, the electrode is subject to "salt error," because the activity coefficient ratio of quinone and hydroquinone ions is related to ionic strength.

Glass electrodes, consequently, seem to be the only ones that hold any promise for *in situ* use. A glass electrode's operation is based on the fact that when a reference electrode and a glass electrode are immersed in the water, a potential difference is set up between them. The half-cell potential of the glass electrode is a function of the hydrogen ion concentration of the sea water (the pH). Thus, by maintaining a constant potential across the reference electrode, the potential across the glass electrode becomes directly related to the pH of the sea water.

It is somewhat difficult to convert this potential to a usable sensor output because the output impedance is on the order of 10^{17} ohms and the output voltage is between 50 and 60 millivolts.

Glass electrodes have not been developed to a point of extremely high accuracy. At the present time, instruments capable of accuracies on the order of ± 0.1 pH units are available. However, if these devices were operated from buoys, fouling problems would probably be so great that this accuracy would not even be approached. Just as with the dissolved oxygen electrode, some way of periodically cleaning the electrode thoroughly must be devised.

907. State of the Art

The state of the art for *in situ* chemical measurements is still in its infancy. The largest problems seem to be instrument drift with time and the deleterious effects of fouling. These are difficulties that, to a certain extent, beset all oceanographic instruments, but they are particularly annoying with regard to those designed for measuring chemical parameters.

Sources and Additional Reading

Ben-Yaakov, S. "A Method for Calculating the In Situ pH of Seawater." *Limnol. Oceanogr.*, 15(2), 1970.

Ben-Yaakov, S., and I. R. Kaplan. "pH Temperature Profiles in Ocean and Lakes Using an In Situ Probe." *Inst. of Geophysical and Planetary Physics*, Pub. 670, Univ. of Calif., Los Angeles.

Bruevich, S. V. "Contemporary Basis of Electromagnetic and Colorimetric Determinations of pH in Sea Water." (In Russian). *Trudy Inst. Okeanol., Akad. Nauk, SSSR,* 47, 1961.

Carritt, D. E., and J. H. Carpenter. "Comparison and Evaluation of Currently Employed Modifications of the Winkler Method for Determining Dissolved Oxygen in Sea Water: a NASCO Report." *J. Mar. Res.*, 24(3), 1966.

Duxbury, A. C. "Calibration and Use of a Galvanic-Type Oxygen Electrode in Field Work." *Limnol. Oceanogr.*, 8(4), 1963.

Fonselius, S. H. "The Oxygen Analysis During the Informal Intercalibration Meeting in Copenhagen." *UNESCO, Techn. Papers, Mar. Sci.,* No. 3, 1965.

Fremling, G. R., and J. J. Evans. "A Method for Determining the Dissolved-Oxygen Concentration Near the Mud-Water Interface." *Limnol. Oceanogr.,* 8(3), 1963.

Ivanenkov, V. N. "Changes of pH, Alkalinity, P and O_2, Caused by the Inner Surface of Nansen Bottles." (In Russian, English summary). In: *Studies of the Indian Ocean (Third Voyage of the Vityaz), Trudy Inst. Okeanol. Akad. Nauk, SSSR,* 64, 1964.

Kanwisher, J. W. "Oxygen and Carbon Dioxide Instrumentation." *Mar. Sci. Instrum.,* 1, 1962.

Morel, A. "Au sujet du dosage colorimétrique de l'oxygène dissous." *Rapp. Proc. Verb., Réunion, Comm. Int. Exp. Sci., Mer. Méd., Monaco,* 17(3), 1963.

———. "Mise au point d'une méthode spectrophotométrique pour le dosage de l'oxygène dissous dans les eaux de mer." *Bull. Inst. Océanogr., Monaco,* 64(1332), 1965.

Park, K. "Oceanic CO_2 System: an Evaluation of Ten Methods of Investigation." *Limnol. Oceanogr.,* 14(2), 1969.

Park, K., M. Oliphant, and H. Freund. "Conductometric Determination of Alkalinity of Sea Water." *Analyt. Chem.* 35, 1963.

Riley, J. P., and G. Skirrow, eds., *Chemical Oceanography.* Academic Press, 1965.

Rotschi, H. "La détermination de l'oxygène dissous par la méthode de Winkler: une evaluation statistique de differentes sources d'erreurs." *Cah. Océanogr.,* CCOEC, 15(7), 1963.

Rudolf, W. "Eine Methode zur kontinuierlichen Analyse des CO_2 Partialdruckes in Meerwasser." *Meteor Forsch. Ergbn.,* (B) 6, 1971.

Solov'yev, L. G. "A Ship Oxymeter of High Sensitivity." (In Russian; English abstract). *Trudy Inst. Okeanol,* 83, 1967.

Solov'yev, L. G., and A. M. Tsvetkova. "Experience in the Use of the Oxymeter 'IOAS' for Continuous Determinations of Oxygen Forming in the Process of Photosynthesis." (In Russian; English abstract). *Okeanologiia, Akad. Nauk, SSSR,* 6(3), 1966.

Vanlandingham, J. W., and M. W. Greene. "An In Situ Molecular Oxygen Profiler: a Quantitative Evaluation of Performance." *J. Mar. Techn. Soc.,* 5(4), 1971.

Whitfield, M. "A Compact Potentiometric Sensor of Novel Design. In Situ Determination of pH, pS^{2-} and Eh." *Limnol. Oceanogr.,* 16(5), 1971.

Wolfe, E. G., C. E. Cushing, and F. W. Rabe. "An Automated System for Multiple Recording of Diurnal pH." *Limnol. Oceanogr.,* 16(3), 1971.

Won, C. H., and K. S. Park. "The Error in Microdetermination of Dissolved Oxygen Instead of Macrodetermination by Winkler Method." (In Korean; English abstract). *Bull. Pusan Fish. Coll. (Nat. Sci.),* 8(2), 1968.

CHAPTER 10

Measurement of Waves and Tides

1001. Classical Results

Until about 1950 the major thrust of investigators involved with the study of waves was to find solutions to the classical wave equation. By making what seemed to be reasonable assumptions under what might be called reasonable boundary conditions, solutions were found, perhaps the most successful being the one generated by George B. Airy in the middle of the nineteenth century. Airy's solution involved an expression for the wave speed of a sinusoidal profile as given in

$$c^2 = \frac{gL}{2\pi} \tanh \frac{2\pi d}{L} \tag{1}$$

where c = wave speed
 g = acceleration of gravity
 L = wavelength
 d = water depth.

1002. Deep-Water and Shallow-Water Waves

The hyperbolic tangent is a mathematical function that has two very convenient extremes. When the argument of a hyperbolic tangent is small, the hyperbolic tangent is approximately equal to its argument, that is

when $\frac{2\pi d}{L}$ is small

$$\tanh \frac{2\pi d}{L} \simeq \frac{2\pi d}{L} \tag{2}$$

At the other extreme, when the argument of a hyperbolic tangent is large, the value of the hyperbolic tangent is closely approximated by unity. Thus

$$\tanh \frac{2\pi d}{L} \simeq 1 \tag{3}$$

when $\frac{2\pi d}{L}$ is large

131

Since the argument of the hyperbolic tangent in Airy's formula [Equation (1)] involved the ratio of the water depth to the wavelength, it seemed only natural to introduce the concept of the relative depth. The words *shallow* and *deep* are commonly used to describe wave approximations. Thus, a wave in water that is very shallow compared to the length of the wave, that is, $\frac{d}{L}$ is small, is called a *shallow-water wave,* while a wave in water very deep compared to its length, that is, $\frac{d}{L}$ is large, is called a *deep-water wave.*

These mathematical approximations result in two distinct types of water waves whose properties differ markedly. Equation (1), for example, shows that the wave speed of a deep-water wave is given by

$$c_d^2 \simeq \frac{gL}{2\pi} \tag{4}$$

and that of a shallow-water wave is given by

$$c_s^2 \simeq gd \tag{5}$$

Note that deep-water waves have a speed related to their wavelength, while shallow-water waves do not. Deep water is a dispersive medium for surface waves, but shallow water is not.

In addition, the orbital motion produced by deep waves is not at all the same as that produced by shallow waves and, for measuring purposes in particular, it is necessary to note that the pressure response of a column of water to a surface wave is markedly different for a deep- and a shallow-water wave. The amplitude of the pressure variation produced by a deep-water wave is given by

$$(\Delta P_d)_z = \bar{\rho}gAe^{-2\pi z/L} \tag{6}$$

where $(\Delta P_d)_z$ = pressure variation amplitude at depth z produced by a deep-water wave
$\bar{\rho}g$ = average weight density of the water column
A = profile amplitude of the wave at the surface (usually half the wave height).

In contrast to this is the amplitude of a pressure variation produced by a shallow-water wave

$$(\Delta P_s)_z = \bar{\rho}gA \tag{7}$$

where $(\Delta P_s)_z$ = pressure-variation amplitude at depth z produced by a shallow-water wave

The pressure variation produced by a deep-water wave falls off exponentially with the depth, while that produced by a shallow-water wave is independent of the depth. Furthermore, the exponential variation with depth associated with a deep-water wave is a function of the wavelength. In other words, in deep water short waves are attenuated much more rapidly with depth than are long waves, so that the water column tends to filter out waves that have short wavelengths. Since a deep-water wave is in reality a wave in water that is deep compared to the wavelength, in shallow water there are waves that may be classified as deep-water waves because of their extremely short wavelength. Thus, in any depth of water, a pressure sensor has some cutoff frequency caused by this filtering effect of the water column for deep-water waves.

1003. Sea and Swell

The terms *wavelength* and *period* are very often mentioned in attempts to describe water waves, and, since those two parameters are related by the wave speed ($c = L/T$), a wave that has a long wavelength tends to have a long period.

Up to this point, water waves have been discussed not only as being repetitive in nature but as having sinusoidal profiles. In actuality the sea surface is not like this at all. Especially in areas where they are being developed by an active wind, waves are short-crested, short-lived, and appear to be anything but regular and periodic. This relatively chaotic situation is termed *sea*. Once the waves leave the influence of the wind that generated them, the configuration tends to become more predictable. The waves seem to become somewhat longer-crested and somewhat more regular in shape. This type of surface configuration is usually referred to as *swell*. Solutions to the wave equation, such as that of Airy, seem more applicable to the condition of swell than to that of sea.

1004. Wave Spectra

To treat the problem of the chaotic nature of sea waves, investigators approached the condition of sea from a statistical point of view and treated it as a random surface: instead of being primarily concerned with profile, they turned their attention to the energy content of the sea surface, because that seemed to be the most profitable consideration. They introduced the concept of a spectrum wherein a particular sea was noted to have a wide range of periods, each having a somewhat different energy content. Figure 10-1 shows a typical spectrum of waves present in a sea under the influence of a strong wind. Note that both short-period waves and very-long-period waves contain some energy, but most of it is concentrated within a specific band of periods, more or less characteristic of this sea.

The type of spectrum obtained was a function of three parameters; wind speed, wind duration, and fetch (the area over which the wind blows at the same speed and in the same direction for a given period of time). Up to a certain point increasing any of these three parameters results in the sea containing more surface energy in the form of greater wave heights and longer

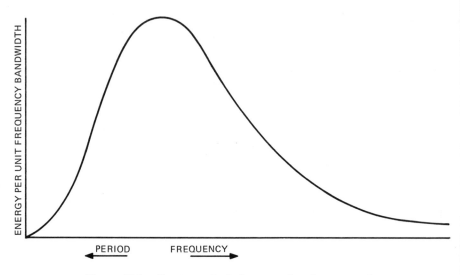

Figure 10-1. Spectrum of wind-generated surface waves for a fully developed sea.

periods. The certain point indicated above is called the *fully developed,* or *fully arisen, sea.* When the wind blows over a water surface, an increasing amount of energy is transferred to the water if fetch, wind speed, or duration is increased. The fully developed sea condition is reached when any additional energy transfer is dissipated in one manner or another. Table 10-1 gives the fetch and duration conditions necessary for a fully developed sea under varying wind speeds.

Table 10-1 Minimum Fetch and Duration to Obtain Fully Developed Sea for Various Wind Speeds

Wind Speed (knots)	10	20	30	40	50
Min. Fetch (nautical miles)	10	75	280	710	1420
Min. Duration (hours)	2.4	10	23	42	69

A fully developed sea spectrum may be stipulated for each wind speed, but a sea that is not fully developed contains a somewhat smaller amount of energy and has a somewhat different spectrum. Figure 10-2 shows a spectrum for a fully developed sea, and the dotted line indicates the spectrum that would be present if the sea were somewhat less than fully developed. Note that the effect is to eliminate some of the longer-period waves, so that, as the sea becomes more and more fully developed, more waves with longer periods are produced.

Since the concept of a spectrum implies that there are waves of all periods within a generating area, classical theory, such as that of Airy, can be used. It may be imagined that a sea is composed of an infinite number of infinitely small amplitude sine waves, each having a different period, wavelength, and direction, and it is the instantaneous sum of these waves that produces the apparently chaotic condition known as *sea*.

However, once these waves leave the generating area and move away from the wind energy source, they travel at their individual speeds, as given by Equation (4). Thus, the longer waves move faster than the shorter waves and, after a while, only long waves will be present, since they will have outstripped the shorter ones. Groups of these long waves are called *swell*.

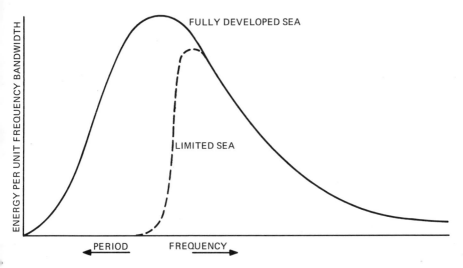

Figure 10-2. A fully developed sea spectrum contrasted with that for a sea whose fetch or whose duration is limited.

1005. Wave Measurements

Wave measurement is not easy. The simple variation of the surface level with time is not the only factor involved: there is also the directional aspect of the wave spectrum that might be produced by a time-dependent record. Generally speaking, most wave energy tends to move in the direction of the generating wind; however, some energy is distributed throughout a ±90° angle on each side of the wind direction. Thus it appears desirable that the time variation of water level be measured at three points, at the very least, in order to allow both the spectrum and its directional characteristics to be determined. Of course, if it were possible to measure the spectrum directly, that would be even better.

Wind-generated waves in the open ocean have statistically significant amounts of energy within periods ranging from a couple of seconds up to about 20 seconds. Although there are wind waves that have periods greater than 20 seconds, the amount of energy contained in most of them is so small that it may be disregarded.

Wave heights vary greatly, the biggest wave ever measured being 112 feet high, but such a height is unusual. Waves running more than 30 or 40 feet high constitute what is considered an extremely high sea. In the Anarctic, where there are long fetches and durations, waves are markedly higher than they are in the region of the equator, where average wind speeds are somewhat less. Consequently, the maximum amplitude capabilities of an instrument should be determined by the area in which the instrument is to be used.

1006. Long Waves

Waves of periods longer than indicated above are present in the ocean, but they are generated by forces other than the wind. Two types of waves have periods longer than 20 seconds; tides and seismic sea waves, the latter being commonly called *tsunami,* or *tidal waves.* Tides have major period components between 12 hours and 24 hours, 50 minutes, and height ranges from less than one foot up to 35 or 40 feet. However, again, the design of the measuring instrument depends upon the area in which it is to be used, because tidal ranges vary markedly from place to place.

Seismic sea waves are generated by underwater earthquakes and, since they occasionally do a great deal of damage in coastal areas, they are of some interest. These waves usually have periods in the 10-to-20-minute range and, at sea, their height is on the order of magnitude of a few feet; however, in coastal areas, they sometimes build to over 20 feet in height. Since their periods are about two orders of magnitude greater than those of wind-generated waves, tsunami are usually measured with tide gages.

1007. Internal Waves

Besides waves that are generated at the water surface, occasionally waves exist within the water column, usually at the interface between two water masses, such as a thermocline. Waves existing at such an internal interface are called *internal waves,* or *interfacial waves,* and typically their periods are somewhere in the range of 1 to 20 minutes. They sometimes reach over 100 feet in height. The exact causes of these waves are not known; however, surges generated by moving weather systems, tide, or subsurface currents might serve as energy sources. It does not require too energetic a natural phenomenon to produce waves of this type, because of the small difference in density between the two fluids involved.

1008. Measuring Internal Waves

Internal waves can be measured either by determining the temporal variation of temperature at the average interface depth, or by determining the vertical excursion of a neutrally buoyant float placed at the interface.

The first determination is usually made with a string of thermistor beads or a thermistor chain which continuously monitors the temperature at a number of discrete depths and allows the internal wave structure to be determined at many depths simultaneously. This type of device can either be deployed from a fixed position, such as a tower or buoy, or it can be towed by a vessel. An 800-foot thermistor chain towed by the USS *Marysville* gave a two-dimensional picture of oceanic thermal structure and internal waves all the way from the surface to a depth of 800 feet.

The second determination is made by means of a feedback circuit. The tension in the suspension cable is kept constant by using the feedback signal either to pull in or pay out cable, and this vertical motion is recorded. The result is a plot of the variation of the depth of the density interface as a function of time, which is the profile of the internal wave.

A rough idea of the presence of an internal wave can be determined from a mechanical bathythermograph cast. If a BT slide shows two different temperature-*vs*-depth traces in the region of the thermocline, the chances are pretty good that this dual trace is the result of an internal wave. A difference in the depth level of the isotherm when the instrument was lowered and the level when it was raised could be accounted for by the presence of an internal wave (*see* Figure 10-3).

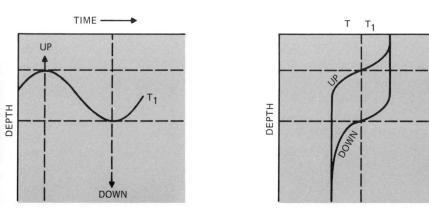

Figure 10-3. A double bathythermograph trace produced by an internal wave.

1009. Pressure-Sensor Wave Meters

Perhaps the easiest way to measure surface waves is with a pressure sensor placed on the bottom in relatively shallow water. From an analysis of the pressure variation as a function of time, the surface profile can be extrapolated. All the pressure sensors discussed in Chapter 3 have been used at one time or another by many different investigators. Since it is desirable to get as much sensitivity as possible, an attempt is usually made to measure the differential, rather than the absolute, pressure because absolute pressure at depth is so large compared to the variations being measured.

A device that measures differential-pressure output directly consists of two oil-filled chambers connected to one another by a capillary tube and bridged by a strain-gage differential-pressure transducer that exposes one end of the gage to each chamber. One chamber is exposed to ambient pressure, so that very rapid pressure changes directly affect one side of the transducer but reach the other side in a greatly attenuated form, after passing through the hydraulic filter. The resultant pressure differential activates the transducer. Very slow pressure changes are negligibly affected by the filter, because the fluid has time to pass through the capillary tube from one chamber to another. Therefore, by choosing suitably the diameter of the capillary tube, it is possible to design a device whose frequency characteristics are known.

Another way of obtaining differential pressure is to use ambient bottom pressure as the reference pressure. This is done by leaving open both sides of a differential-pressure gage as the device is lowered to the bottom. When the gage reaches the bottom, a solenoid valve seals the reference chamber, permitting the sensor to operate at full sensitivity in the presence of high ambient pressure.

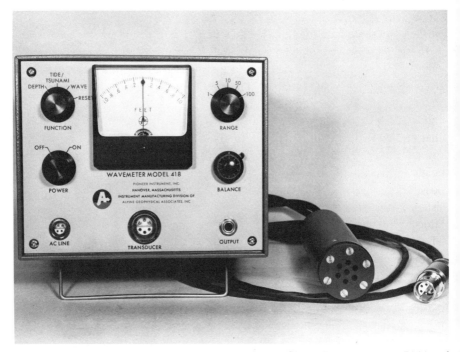

Pioneer Instrument, Inc., a Division of Alpine Geophysical Associates, Inc.

Figure 10-4. A wave meter.

The output of the sensor may be a variation in resistance or a variation in voltage, depending upon the device used to detect pressure changes. One system uses a pressure-sensitive inductance as one arm of an AC bridge driven by a one-kilohertz oscillator. A change in pressure applied to the pressure-sensitive element varies the AC output of the bridge.

1010. Other Wave Meters

In addition to pressure variations, acoustic energy return is also used as the measurand for wave heights. The configuration is that of an upside-down echo sounder in which a sound beam is transmitted toward the surface and reflected back to the receiver on the bottom. The disadvantage in a system of this nature is that the beam has a finite spreading angle and therefore tends to average out the small waves appearing in the surface profile.

An obvious way to measure waves in shallow water is to implant some sort of staff into the bottom and measure the variation of water level about the staff. This can be done in many ways, some of which involve the closing of reed switches, the varying of the staff resistance, tuned transmission lines, and capacitance variation.

In the reed-switch staff, a series of reed switches is used to connect various resistances into a circuit in response to a doughnut-shaped magnetic float that follows the wave profile. A fixed-source voltage is used in the resistive network, so that the output voltage is a function of the wave height.

In a resistor-wave staff, the sensor staff is composed of a number of equal-value resistors placed in series at selected intervals along the staff. The staff is constructed in such a manner that only contact points are exposed to sea water in between each of the resistors. Thus, as the sea water rises, more and more resistive elements are shorted out, and the result is a step-by-step change in resistance proportional to the wave height. As above, a fixed-source voltage is used to obtain a voltage output directly related to wave height. Instead of using discrete resistances to get a stepped output, it is possible to design a resistance wire that gives a continual variation of resistance with water level. This resistance wire is often part of an RC oscillator circuit that changes frequency in accordance with resistance variations. Thus, the output is a variation in frequency rather than in amplitude.

In the case of tuned transmission lines, two stainless-steel wire ropes are held parallel by tension sufficient to resist wind and wave forces. The device is mounted normal to the sea surface in such a manner that the immersion level changes the length of the exposed tuned pair by shorting out all below the water surface. In this manner, a linear DC output proportional to the water level is obtainable.

The capacitance probe uses the water itself as one plate of a capacitor and a conducting rod as the second plate. An insulating material wound around the rod serves as the dielectric, so that as the water rises around the rod, the capacitance of the system changes. Sometimes the rod forms a portion of an electronic oscillator, so that the output varies in frequency as the water level changes. When securely mounted to the bottom, the staff can measure all period waves; however, if the water is too deep for a rigid implantation, a damping disc can be used to form a floating staff, as shown in Figure 10-5.

Figure 10-5. A wave staff with a damping disc.

If the damping disc is placed in a depth greater than half the wavelength of the longest wave to be measured, the staff can respond only to waves whose periods are greater than those of interest. In this manner, only those frequencies being measured will cause a water-level variation around the wave staff. The wave staff is occasionally mounted on the bow of a large vessel and, if the vessel is very much longer than the waves, a good approximation of the wave motion can be obtained.

The ideal situation, however, is somehow or other to measure the motion of the vessel in an absolute frame of reference and compare that measurement with the output of the wave instrument. M. J. Tucker has done this with his ship-mounted wave gear, consisting of pressure sensors mounted below the waterline on each side of the ship. At each pressure sensor an accelerometer is also installed. The signals from the accelerometers are doubly integrated and combined with the signals from the pressure sensor, the theory being that the doubly integrated signal locates the pressure sensor above or below mean-water level. The height of the wave, as measured by the pressure sensor, is then added to give the total wave height.

Of course, accelerometers can be used independently to determine surface profile: an accelerometer is placed on a floating buoy and the output is integrated twice to determine the displacement. Such a device, called Splashnik, was developed by W. Marks and he has used it for some years. It consists of a telemetering accelerometer buoy that can be dropped overboard from a moving ship. The telemetered data are recorded aboard the ship as long as the buoy is within range of the ship. This device is inexpensive enough to be used in an expendable mode.

Any of the devices described above can be used in groups of three or more to obtain directional spectra. However, by using stereo-photographic pairs both power and directional spectra can be determined simultaneously. These photographs can be taken from aircraft, or even from shipboard, if the equipment can be mounted high enough on the mast. The procedure is, by one means or another, to obtain a stereo pair of overlapping photographs of an area. From measurements of each individual wave, the directional and energy spectra can then be determined.

Modern sampling techniques also indicate that it should soon be possible to display the real-time spectrum by proper data processing. When this has been achieved, it will be possible to observe the change in spectrum as a sea develops. This has been done partially by D. Wilson at the Chesapeake Bay Institute of The Johns Hopkins University.

1011. Tide Gages

In contrast to the diversity of devices designed for measuring wind waves, tide gages for measuring much-longer-period waves are all based on essentially the same principle. A tide gage is a long vertical tube sealed at one end, except for a very small opening on the bottom that allows water to enter and leave only very slowly. In this manner, the response of this device can be limited to waves whose periods are on the order of magnitude of minutes rather than seconds, and the high-frequency cutoff can be determined by the size of the entrance hole. The water-level variation is recorded by means of a float mechanically attached to a system of levers driving a chart pen, or it is converted into a digital output.

1012. Measuring Waves from Satellites

Modern microwave techniques make it possible to measure small differences in height from great distances. Thus, not only can water-level variations, such as those produced by tides and tsunami, be measured from satellites, but so can wind-wave heights. A combination of microwave, visible photography, and, perhaps, lasers should make it possible in the not-too-distant future to measure all the wind waves on the world's oceans almost simultaneously.

It seems that a small extension of present techniques would give accuracies on the order of magnitude of ±0.5 meters. If this is so, it might be possible, by applying wave-height and occurrence data from the world's oceans, to develop a worldwide anemometer. Microwave data might also be used with visual photographs showing whitecap areas even further to enhance the information level.

1013. State of the Art

At the present time, under ideal conditions, wave-height data accurate to ±0.1 meters can be obtained with the devices described above. There are many problems, however, connected with the response of these various devices in terms of their output as a function of frequency. Some devices give slightly erroneous water levels, as the speed with which the water level changes varies. In addition, fouling can be particularly bothersome in most of these devices if they are left unattended for any length of time.

Just as it was difficult to design a motion-measuring device that would be satisfactory at both ends of the spectrum, that is, for both small and large speeds, it is difficult to design a device that can measure very small and very large waves simultaneously. Yet information about both of these wave families must be available, if physical phenomena such as those that exist inside the whitecaps of breaking waves are to be studied.

Sources and Additional Reading

Ayers, R. A., and D. J. Cretzler. "A Resistance Wire Water Level Measurement System." *Mar. Sci. Instrum.*, 2, 1963.

Banwell, T. J. "The TCB Wave Gage." *Symposium, Ocean Sci. and Ocean Eng.* Mar. Techn. Soc., Am. Soc. Limnol. Oceanogr., 2, 1965.

Bigelow, J. W. "Low-Cost Wave Recording Systems." *Mar. Sci. Instrum.*, 4, 1968.

Canadian Hydrographic Service. " 'Ottboro' Tide Gauge." *Int. Hydrogr. Rev.*, 41(2), 1964.

De Blok, J. W. "An Apparatus to Generate Tidal Fluctuations and a Modification to Render Constant Flow." *Netherlands J. Sea Res.*, 2(2), 1964.

Farmer, H. G., and D. D. Ketchum. "An Instrumentation System for Wave Measurements, Recording, and Analysis." *Proc. 7th Conf. Coastal Eng., Council on Wave Research,* Univ. of Calif., 1960.

Furuhata, T. "New Electric-Ocean Wave Recorder, MR-Mark III, for the Coastal Wave Stations." *J. Oceanogr. Soc.,* Japan, 18(3), 1962.

Goodheart, A. J., C. W. Iseley, and S. D. Hicks. "Deep-Sea Tide Gage." *Symposium, Ocean Sci. and Ocean Eng.* Mar. Techn. Soc. Limnol. Oceanogr., 1, 1965.

Greenwood, J. A., A. Nathan, G. Neumann, W. J. Pierson, F. C. Jackson, and T. E. Pease. "Oceanographic Applications of Radar Altimetry from a Space Craft." *Remote Sensing of Environment,* 1, 1969.

Hamon, B. V., "Australian Tide Recorders." *Austr. Comm. Sci. Ind. Res. Organ., Div. Fish. Ocean.,* Tech. Pap. No. 15, 1963.

Harris, M. J., and M. J. Tucker. "A Pressure Recorder for Measuring Sea Waves." *Instrum. Pract.,* 17, 1963.

Hasselmann, K. "Determination of Ocean Wave Spectra from Doppler Radio Return from the Sea Surface." *Nature,* Lond., 229(1), 1971.

Kaplan, P., and D. Ross. "Comparative Performance of Wave Measuring Systems Mounted on Ships in Motion at Sea." *Mar. Sci. Instrum.,* 4, 1968.

Kawashima, R. "On the Measurement of Ocean Waves. 1. A Telemetering System for the Measurement of Ocean Waves by Means of the Use of a Bamboo Stick Wave Pole." *Bull. Fac. Fish.,* Hokkaido Univ., 14(1), 1963.

Kolesnikov, A. G., and V. V. Efimov. "Apparatuses for the Measurement of Energy Transmitted by Normal Wind Pressure upon Sea Waves." (In Russian). *Okeanol., Akad. Nauk, SSSR,* 4(3), 1964.

Kozyrev, M. A., V. P. Liverdi, and N. I. Ryzhkova. "Method of Determining Wave Elements from a Ship by Two Spaced Float-Type Recorders." (In Russian). *Trudy Morsk. Gidrofiz. Inst.,* 26, 1962 Transl.: Scripta Tecnica, 26, 1964.

———. "Performance of the Photographic Slit Wave Meter in Conjunction with a Vertical Movement Meter." (In Russian). *Trudy Morsk. Gidrofiz. Inst.* 26, 1962. Transl.: Scripta Tecnica, 1964.

Kronengold, M., J. M. Loewenstein, and G. A. Berman. "Sensors for the Observation of Wave Height and Wind Direction." *Mar. Sci. Instrum.,* 3, 1963.

Kukosik, S. J., and C. E. Grosch. "Pressure-Velocity Correlations in Ocean Swell." *J. Geophys. Res.,* 68, 1963.

LaFond, E. C. "Internal Waves and Their Measurement." *Mar. Sci. Instrum.,* 1, 1962.

Peep, M., and R. J. Flower. "The Chesapeake Bay Institute Wave Follower." *The Johns Hopkins Univ. Tech. Rept.* 58 (Ref. 69–11), 1969.

Pigeon, N. D., and W. W. Denner. "A Resistance Tide Gauge." *Mar. Sci. Instrum.,* 4, 1968.

Mark, R. B. "Shipboard Ultrasonic Wave Height Sensor." *Mar. Sci. Instrum.,* 2, 1963.

Marks, W., and R. G. Tuckerman. "A Telemetering Accelerometer Wave Buoy." In *Ocean Wave Spectra.* Prentice-Hall, Englewood Cliffs, 1963.

Roumégoux, L. "Continuous Air Flow Tide Gauge." *Int. Hydrogr. Rev.,* 41(1), 1964.

Shipley, A. M. "On Measuring Long Waves with a Tide Gauge." *Deutsche Hydrogr. Z.,* 16(3), 1963.

Stass, I. I. "Deep-Water Tide Gauge." (In Russian). *Trudy Morsk. Gidrofiz. Inst.,* 26, 1962. Transl.: Scripta Technica, 26, 1964.

Summers, H. J., H. D. Palmer, and D. O. Cook. "Some Simple Devices for the Study of Wave-Induced Surges." *J. Sedim. Petrol.,* 41(3), 1971.

Tucker, M. J. "A Ship-Borne Wave Recorder." *Trans. Inst. Naval Arch.,* 98, 1956.

Upham, S. H. "Electric Wave Staff (Hydrographic Office Model Mark I)." *U. S. Naval Hydrographic Office, Tech. Rpt. 9,* Washington, D. C., 1955.

Volkov, W. G., W. V. Karpovich, and A. P. Kestner. "A Broad Bandwidth Resistance Wave Recorder, Methods of Measurements and Statistical Analysis of Sea Surface Waves." (In Russian; English abstract). *Fisika Atmosfer. Okean., Izv. Akad. Nauk, SSSR,* 3(9), 1967.

Williams, L. C. "An Ocean Wave Direction Gage." *Mar. Sci. Instrum.,* 3, 1965.

Zamaraev, B. D., and A. I. Kalmykov. "On the Possibility of the Determination of the Spatial Structure of the Wavy Surface by Radar Methods." (In Russian). *Fisika Atmosfer. Okean. Izv. Akad. Nauk, SSSR,* 5(1), 1969.

Zetler, B. D., and G. A. Maul. "Precision Requirements for a Spacecraft Tide Program." *J. Geophys. Res.,* 76(27), 1971.

CHAPTER 11

Geophysical Measurements

1101. Gravity and Magnetic Measurements

The basic tools of the marine geophysicist are almost identical with those used by his land-based counterpart. Consequently, the measurement of crustal-sound velocities, gravity anomalies, and magnetic anomalies occupies the major portion of any marine geophysical program. In a previous chapter we discussed the use of sound in the ocean and in the oceanic crust to determine sound speeds and therefore composition of crustal material (*see* Section 809). In this chapter, therefore, we shall confine ourselves to a discussion of gravity and magnetic measurements.

1102. Acceleration of Gravity

Acceleration of gravity is primarily the result of the gravitational attraction of the earth for any body within its gravitational field. Since this gravitational attraction is related to the distance of the body from the center of the earth, any change in elevation affects its value. At greater depths in the ocean, gravitational attraction is somewhat greater than it is at the sea surface. Similarly, gravitational attraction increases in the vicinity of a large mass of high-density material.

Besides its relationship to vertical location and crustal composition, the acceleration of gravity is related to latitude. This turns out to be a result of the fact that acceleration of gravity is, in reality, the result not only of the gravitational force but also of the centrifugal force produced by the rotation of the earth, as is shown in Figure 11-1. Since, at high latitudes, the distance from the earth's axis to its surface is less than it is at low latitudes, the centrifugal effect is greatest at the equator. For this reason, the earth has a somewhat oblate shape and we find that acceleration of gravity is least at the equator and greatest at the poles.

Thus, knowledge of the earth's rotation, size, and crustal composition makes it possible to draw conclusions as to the distribution of gravitational acceleration values over the earth's surface. Any gravity measurements that deviate from these theoretical values are called *gravity anomalies*.

At the earth's surface, average gravitational acceleration is about 980 cm-sec^{-2}. Variations from this value are usually small and, in order to describe them, the gal unit is introduced, one gal being equivalent to an acceleration of one cm-sec^{-2}.

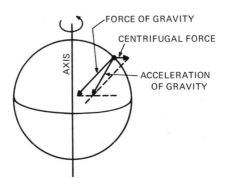

Figure 11-1. Acceleration of gravity is a result of gravitational attraction and the earth's rotation.

1103. Gravity-Measuring Methods

The classical instrument for measuring gravity is a swinging pendulum: the period of the pendulum is measured and, from this, the acceleration of gravity can be determined. However, the Worden gravimeter is much more accurate and it is not subject to the motions of unstable platforms, such as may be experienced at sea. The heart of a Worden gravimeter is a small mass kept in equilibrium by springs. Any deviation from the standard value for acceleration of gravity causes the balance of the system to be slightly upset and the mass to move; it moves down if the gravity anomaly is positive, and up if it is negative. By measuring this movement, the gravity anomaly can be determined directly, whereas the determination is indirect when the total acceleration of gravity is ascertained (*see* Figure 11-2). Seagoing gravity meters of this design have accuracies on the order of magnitude of ±0.02 milligals and can, therefore, detect very small deviations from theoretical expected gravitational acceleration values.

1104. Earth's Magnetic Field

To a first approximation, the magnetic field around the earth is that of a simple dipole: one pole near the northern extremity of the earth's rotational axis and the other near its southern extremity. Examination of this magnetic field at any place on the surface of the earth shows that, at some place in the vicinity of the equator, the field lines are essentially parallel to the earth's

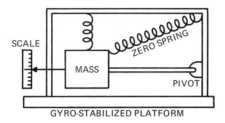

Figure 11-2. Schematic representation of a Worden gravimeter. Any change in acceleration of gravity causes the pointer to move either up or down.

surface. Therefore, the field has no vertical component. However, in the middle latitudes, the earth's magnetic field has both a horizontal and a vertical component, while at the magnetic poles there is only a vertical component (*see* Figure 11-3). The average magnitude of the earth's vertical magnetic field in middle latitudes is about 50,000 gammas, equivalent to 0.5 Oersteds.

The magnetic field of the earth is determined by the magnetic effect of processes active within the inner and outer cores of the earth, as modified by the irregular location of magnetic material within the earth's crust. Thus, on the basis of a simple dipolar model, earth's magnetic field at any given place can be predicted, but the chances are that the actual measured value will be somewhat different. The difference between the predicted value and the actual (normal) value is called a *magnetic anomaly,* and it is this magnetic anomaly that the geophysicist attempts to measure.

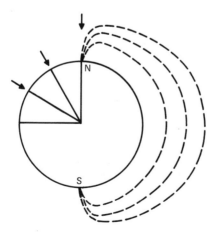

Figure 11-3. The vertical component of the earth's magnetic field varies with latitude.

1105. Measuring Magnetic Anomalies

At one time, the most popular device for measuring magnetic anomaly was the flux gate magnetometer. Although it has in large measure been replaced in recent years by the nuclear precession magnetometer, let us briefly discuss the flux gate unit, since it is used by the U.S. Navy in its magnetic anomaly detection (MAD) gear for submarine detection.

A flux gate magnetometer uses two coils wound so that, in a zero magnetic field, a current passed through them will cause no unbalance if they are in two arms of a bridge. However, when these two coils are placed in an ambient magnetic field they are so oriented that the field tends to add to the magnetic field of one coil while subtracting from the magnetic field of the other. Thus, when a current is passed through them, the coils are no longer in balance under this condition, and their imbalance is a function of the magnetic-field intensity being measured.

What is usually done is to set up a balance in the normal ambient magnetic field so that, in the magnetic field being measured, any deviation from this norm shows up as an anomaly.

A nuclear precession magnetometer offers somewhat greater accuracy than the flux gate, but its signal is somewhat smaller in amplitude. In addition, it has the advantage of producing a frequency-varying signal which, as indicated previously, is more suitable for recording on magnetic tape than is one that varies in amplitude.

The principle on which a nuclear precession magnetometer is based involves the application of a magnetic field to a liquid containing hydrogen (kerosene or distilled water, for example) until a large number of the hydrogen nuclei are given a magnetic moment. When this artificial field is removed, the liquid containing the hydrogen nuclei is placed in the earth's magnetic field whose orientation is different from that of the artificial field. Because of their newly acquired magnetic moment, the nuclei precess in the earth's magnetic field at a frequency determined by the magnitude of the field. A measure of the magnetic field of the earth is then obtained by simply measuring the frequency at which the hydrogen nuclei precess.

Either one of these instruments can measure magnetic anomalies to an accuracy of about ±one gamma, and, under ideal conditions, a tenth of a gamma resolution can be achieved.

1106. State of the Art

In recent years, both gravimeters and magnetometers have decreased in size and, consequently, have been used more and more both at sea and in the air. This trend will probably continue as increasingly comprehensive geophysical surveys over wider areas will be made with greater frequency.

Sources and Additional Reading

Lawrence, L. G. *Electronics in Oceanography.* Howard W. Sams & Co., Indianapolis, 1967.

Nelson, J. H. L., L. Hurwitz, and D. G. Knapp. *Magnetism of the Earth.* ESSA Coast and Geodetic Survey, Pub. 40–1, U.S. Government Printing Office, Washington, D.C., 1962.

Strahler, A. N. *The Earth Sciences.* 2nd ed. Harper & Row, New York, 1971.

Woollard, G. P., and J. C. Rose. *International Gravity Measurements.* Society Exploration Geophysicists, Spec. Pub., 1963.

———. "Gravity." In *Research in Geophysics.* Vol. 2. H. Odishaw, ed. MIT Press, Cambridge, 1964.

CHAPTER 12

Instrument Platforms

1201. Fixed and Moveable Platforms

In the previous chapters we have discussed a number of devices for measuring various oceanographic parameters. In all cases, for these devices to be used, it is necessary to take them to sea. The methods by which they are taken to sea and deployed, i.e., the various platforms in common usage for the implementation of any program, are the subject of this chapter.

Some platforms are fixed and some are moveable. Fixed platforms collect continuous data at one place, while moveable platforms collect data at many places. Buoys, lightships, towers, and dockside laboratories are fixed platforms; ships, aircraft, and satellites are moveable platforms.

Fixed platforms are used either singly to acquire long-term information about a specific location, or in groups to attempt synoptic coverage of an entire area.

Moveable platforms also are used to acquire snyoptic data, but when this is done the assumption is made that the environment does not change appreciably during the course of data-taking. This assumption may be invalid in the case of a conventional surface vessel, although in many parts of the ocean changes with respect to time are so small that the assumption does remain valid. Often, when large areas are to be closely studied, multiple moveable platforms are used simultaneously in order to reach a closer approach to true synoptic measurements. Since the main objective of any scientific endeavor is the prediction of events, the most desirable way of acquiring data is synoptically: this is because data so acquired can be used in various forecasting mathematical models. However, many oceanographic and atmospheric phenomena can be examined with data acquired at specific locations, and for these a fixed platform is quite suitable.

1202. The Ideal Platform

In all cases platform designers do their best to approach the condition of the ideal platform. The ideal platform is one that does not, through its motion or through any other of its characteristics, affect the quality of the data produced from it. Thus, an ideal platform allows any instrument to be used

Figure 12-1. Argus Island: an example of a fixed platform.

as if it were operated in a laboratory. However, in many cases this ideal specification is not even closely obtainable, and data analysis has to allow for the shortcomings.

It appears that, first and foremost, a platform must be stable. It must be fixed with respect to some desired reference level, which could be either an arbitrary geopotential surface or the undulating surface of the sea. For example, for measuring the wind speed at a constant height above the ocean surface, the ideal platform is one that exactly follows the naviface and keeps the wind sensor at a constant level above this varying surface.

On the other hand, for measuring the water temperature at a depth of one hundred meters (measured from average sea level), it is undesirable to have the sensor bobbing up and down with the surface water and changing its vertical position. So, what is ideal in a platform depends on the use to which the data is to be put and the design of the sensor itself.

For much oceanographic work, especially that of a surveying nature, the ideal platform is stable with respect to a fixed set of coordinates and large enough to carry the necessary high-quality power supplies that form an integral part of the instrument systems to be used.

1203. Buoys

Although other fixed platforms are used, the most common are buoys, since they are somewhat less expensive and less complex than either lightships or permanently installed towers. Buoys are classified as either surface-following or stable, depending on the reference coordinates chosen.

Surface-following buoys, most of whose structure is relatively close to the surface and whose mass is distributed horizontally, respond closely to the configuration of the ocean surface produced by sea or swell. These buoys take many forms, but discs, toroids, catamarans, spheres, and paraboloids have been used most successfully. Of all the forms tried to date, paraboloids seem to move up and down with the changing configuration of the sea surface with the smallest amount of tilt. None of these configurations, however, work too well in seas generated by winds of hurricane speed, even though surface-following buoys are used for measuring wind speed at a constant level above the naviface.

Stable buoys do not readily respond to surface-wave effects. Their mass, which is distributed vertically, may extend more than a hundred meters below the water surface. Examples of this type are the Sea-going Platform for Acoustic Research (SPAR) and the Floating Instrument Platform (FLIP). Both of these spar-type buoys are about 108 meters long, the major difference between them being that FLIP is a manned buoy and SPAR is not. In a horizontal position, they are surface-towed to their destination, then flooded, one end sinking to a vertical attitude. When it is desired to move one of these buoys, the water is pumped out and the device once more floats in a horizontal attitude.

Spar buoys are certainly more stable than surface-following buoys, in the sense that they remain in a fixed geometrical location. However, they are quite a bit bulkier, harder to manage, and therefore more expensive than surface-following types.

The moorings used for buoys fall into two general classes: fixed moorings and compliant moorings. Fixed moorings allow relatively little displacement of the buoy by wind or current, while compliant moorings allow substantial horizontal motion. Free-floating buoys are extreme examples of compliant moorings.

One type of fixed mooring is the taut-line moor, which consists of an anchor with a taut line connected to the buoy. Since the line is taut to the bottom, any horizontal motion of the buoy is accompanied by a vertical displacement, so that, in heavy seas or strong currents, there is a tendency for the buoy to be swamped, as shown in Figure 12-2. This tendency to swamp may be offset by a somewhat more compliant mooring, such as the one-point mooring. Here, a taut line is attached to a flotation buoy and, from this float, a horizontal line is attached to the buoy of interest (*see* Figure 12-3). Thus, the float may be swamped but the buoy itself remains at the surface and maintains its position within the length of the line attached to the float. A two-point mooring (*see* Figure 12-4) has two taut lines used with two floats, between which the buoy of interest is attached with two horizontal lines. This arrangement allows somewhat less horizontal motion than does a one-point mooring, and makes it possible for some of the advantages of both fixed and compliant moorings to be achieved. For even more desirable characteristics, a three- or even a four-point mooring may be attempted.

Once the buoy configuration has been chosen and its type of mooring determined it is necessary to settle on the instrumentation package to be carried aboard the buoy, along with the power supply needed to drive the instruments. If the buoy is to be untended for long periods of time, it might be necessary to put aboard a small power source, such as a diesel generator, to drive the instrument package. However, if the power requirements are small, some combination of a solar cell and storage battery might suffice. The

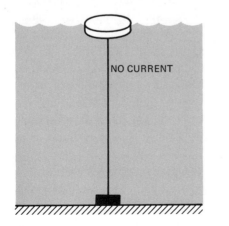
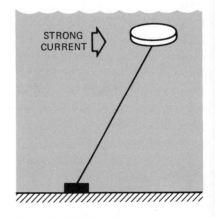

Figure 12-2. A taut-line buoy swamped by a strong current.

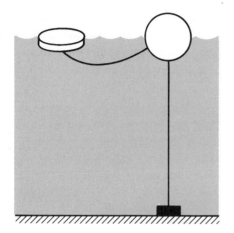
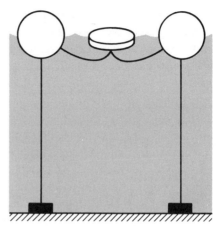

Figure 12-3. A one-point mooring.   **Figure 12-4. A two-point mooring.**

power requirement may also be a factor in determining the type of sampling system used. If a continuous depth profile of certain parameters is desired, the sensor package has to be programmed to move up and down, and that requires a winch, which causes a heavy drain on the power system. On the other hand, if measurements at discrete depths are satisfactory, a number of sensor packages hung at various depths on a cable will obviate the need for a winch and therefore make the power requirements somewhat less.

How the data collected are to be transmitted is an important facet of design, since it must be determined whether they are to be telemetered to some collection station or stored in some manner and picked up at selected time intervals. (*See* Chapter 13 for discussion of the transmission and analysis of data.)

Both systems have advantages. A data-storage system, wherein there is no telemetry requirement, allows greater motion of the buoy package, because there is no need to keep a radio antenna oriented for data transmission at all times. However, ships have to go out and service the buoys at frequent intervals. The expense of this operation may very well offset the savings accomplished by specification of a buoy in contrast to a survey vessel.

1204. Surface Ships

The most common moveable platform is the oceanographic survey ship. This is a surface vessel designed to move as rapidly as possible from one geographical location to another at which it lowers oceanographic measuring instruments to obtain oceanographic data. Many of these ships are converted tenders and other vessels that have outlived their usefulness in their original configurations. The modern trend, however, is to design and build ships from the keel up for oceanographic research. One of the objectives of the designers of these vessels is to make them as stable as possible and, for this reason, anti-roll tanks are usually installed. In the effort to achieve maximum stability, catamaran hulls have been used, but it has not been established that double-hulled vessels are superior to single-hulled ones in a hove-to configuration.

Naval Research Laboratory

Figure 12-5. Oceanographic research ship USS Hayes (AGOR-16).

In addition to the stability of the platform, the designers of oceanographic vessels are faced with unique problems in attempting to provide high-capacity electrical generators, large versatile winches, mechanical and electrical equipment that produces minimal noise and vibration, adequate laboratory space, and long cruising ranges. Modern oceanographic vessels also have to have air-conditioned facilities, primarily to protect the integrity of their laboratory equipment, but also to enable scientists to accomplish more and better work during the summer months and in hot climates.

1205. Submersibles

To solve the problem of instability in terms of motion produced by the varying sea surface, it has been suggested that subsurface vessels be used for oceanographic research, and a few submersibles have been employed for this purpose.

A submersible has the distinct advantage of operating in an environment much less active than the naviface, and is, therefore, more suited for such delicate measurements as the acceleration of gravity. Furthermore, a scientist in a submersible can actually get into the environment he is studying. Biological and geological oceanographers especially benefit from this type of vessel, because they can see the features that are the objects of their study.

Figure 12-6. Inside the Trieste II.

However, for physical and chemical oceanographers it is not particularly beneficial to work from submersibles. It is certainly just as difficult to measure salinity and temperature variations from a submersible as it is from a surface vessel; these and other parameters cannot be measured by the human eye. Disadvantages of submersibles are their high cost, low speed, and, with rare exception, their endurance times and payload capabilities leave much to be desired.

One nuclear submersible has been built for oceanographic research, the NR-1; however, she was built by and for the U.S. Navy and how much actual oceanographic research will be accomplished aboard this highly classified naval vessel is questionable.

1206. Scuba

If biological and geological oceanographers benefit by penetrating the environment aboard a submersible and being able to look out of the ports and see living organisms, bottom composition, and bottom contour *in situ*, it is then just as desirable for these people to use the commonly available scuba diving apparatus as a platform in shallow waters. Using scuba gear as an instrument platform allows not only firsthand observation of interesting phenomena, but the transportation and use of many types of oceanographic instruments that cannot be conveniently deployed on the end of a cable. Perhaps even more important is the capability that is afforded of studying the collecting characteristics of various fish and plankton nets. Scuba gear allows geological oceanographers to examine at leisure some of the rock formations that appear in exposed areas of the sea bottom within the context of the entire formation, rather than as isolated samples.

There is no doubt that scuba gear has added an entirely new dimension to the study of biological and geological oceanography, but it should be emphasized that this is just another tool and not an end in itself.

1207. Satellites

Another recent addition to the flock of platforms available to the oceanographer is the satellite. Satellites as data-transmission links are discussed in Chapter 13. Here, they are discussed as platforms for sensor systems. Electromagnetic energy in its various forms can be used to determine a great deal about the ocean's surface layers from great altitudes. Regular and enhanced visible light, infrared radiation, radio waves, and radar frequencies all can be used to determine various oceanographic parameters. Many satellite sensors are in the process of development, but it is a foregone conclusion that, within a few years, it will be possible to measure such parameters as sea state, erosion, turbidity, hydrography, sea color, productivity, presence and magnitude of currents, ice, tidal changes, and storm surges from satellites. Accuracies may not be as great as those obtainable at the ocean surface, but for many purposes they will be satisfactory.

There are many advantages in using a satellite as a sensor platform, the greatest of which is that the satellite's relatively short period of 90 minutes provides repeated coverage of all areas. Thus, changes in the environment that take place within this small time interval can be detected with great ease. Also, data-acquisition time is reduced because a satellite travels so rapidly and covers such a large area as it sweeps around the earth. Although a satellite is a tremendously expensive platform, it gets so much data in such a short period of time that each piece of information costs less than it would if acquired in any other way. On a dollar-per-data-bit basis, this is undoubtedly the least expensive way of obtaining oceanographic data.

Another advantage of a satellite is the fact that it produces almost simultaneously pictures of the entire world ocean, presenting for the first time an opportunity to come very close to obtaining a synoptic picture of the surface hydrosphere. Since a space instrument platform moves in a relatively uniform environment, it is very stable and the data produced can be expected to reflect this stability. Each pass of a satellite covers a large area, so that small details and their correlation with other small details can be seen in the same swing of the satellite, even though the phenomena may be separated by some distance.

A satellite is also flexible in the sense that, by proper choice of orbit, it may be used as either a moveable or a fixed platform. If the orbit is conventional, that is, it lasts about 90 minutes, the platform can be considered moveable. However, as has been demonstrated with communications satellites, it is possible to put a satellite up so that it will rotate with the same angular speed as does the earth itself. In this manner, the satellite appears to occupy the same position in the sky. Thus, is can be considered a fixed platform in that it can continuously monitor a specific oceanic area. In this configuration it is obviously similar to a fixed surface oceanographic platform.

The major disadvantage of using satellites as sensor platforms is that the electromagnetic energy used in their sensors does not penetrate the water to any great depth. In the middle of the electromagnetic window (the blue-green portion of the visible spectrum), for example, penetration on the order of 100 meters is the most that can be expected. Satellite measurements sample only the upper layer of the oceanic environment, so that the third dimension of the ocean has to be measured from buoys, ships, or other platforms of a more conventional nature.

1208. Aircraft

Of course, any measurements that can be made from satellites can be made just as well, if not better, from aircraft. At the present time, infrared and photographic measurements are being made from aircraft on a regular basis, and this practice will continue. Aircraft are also used for magnetic and gravity surveys, for which purposes they have the great advantage of covering a large area in a relatively short period of time.

In recent years aircraft have been used even to obtain conventional data with standard over-the-side sensors and also to collect water samples. Helicopters and airships are obviously well suited for this type of operation but, by means of adroit maneuvering and ingenious sampling techniques, fixed-wing aircraft have also proved to be capable platforms. Since even a small plane can cover a large area in a short time, many oceanographic surveys can be made less expensively from aircraft than from surface vessels.

1209. State of the Art

Of the many platforms mentioned above, the surface ship is certainly the one that supplies most oceanographic data at the present time. Besides oceanographic vessels, many ships of opportunity are being fitted with instrumentation packages so that they can obtain oceanographic data as they go about their usual business of commerce on the high seas.

It is expected that buoys and satellites will, in time, be used to a much

Westinghouse Electric Corporation
Oceanic Division

Figure 12-7. The data-collection buoy to be used in the National Data Buoy Project.

INSTRUMENT PLATFORMS 161

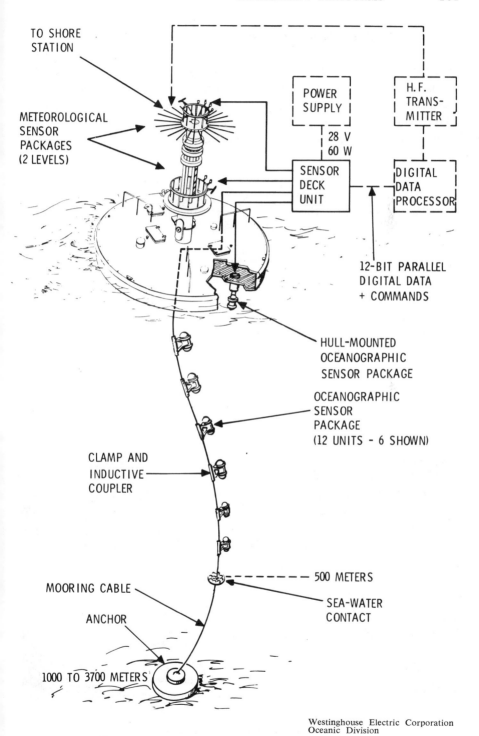

Figure 12-8. A sketch of the buoy shown in Figure 12-7.

greater extent than they are now. The National Data Buoy Project, under the cognizance of NOAA, includes plans for a number of buoys to be placed throughout the world ocean. As may be seen in Figure 12-8, each buoy has a number of meteorological sensors at two heights (5 and 10 meters) above sea level. Oceanographic sensor packages are placed at the surface and at twelve preselected depths down to 500 meters below the surface. Although the parameters measured are not unusual—temperature, depth, conductivity, current, and sound speed in water; and temperature, dew point, air pressure, global radiation, wind speed, and precipitation rate in air—some of the sensors used reflect an advanced state of the art. This is true in the speed sensors particularly. Water motion is measured with a doppler-effect system while the air speed is determined by the use of a vortex shedding unit (*see* Section 614).

The sensors and the power supply, a propane-fueled generator, were both chosen for minimum maintenance. Furthermore, because the system is programmed to transmit data only upon interrogation by a shore station, a major power drain is cut down.

It will be interesting to see whether data from these buoys will supplant data previously received from manned platforms, such as ships.

Sources and Additional Reading

Anonymous. "A Design Report on the Monster Buoy's Sensor Package." *Geo. Mar. Techn.*, 2(6), 1966.

Botzum, J. R. "Nimbus-C Aids Ocean Studies." *Geo. Mar. Techn.*, 2(6), 1966.

Brainard, E. C. III. "Drone Surface Vehicles for Remote Controlled Surveys." *Mar. Sci. Instrum.*, 4, 1968.

Cox, C. S., B. P. Johnson, H. Sandstrom, and J. H. Jones. "A Taut Wire Mooring for Deep Temperature Recordings." *Symposium, Ocean Sci. and Ocean Eng.* Mar. Techn. Soc., Am. Soc. Limnol. Oceanogr., 2, 1965.

Daubin, S. C., "Deep-Sea Buoy Systems." *Instrum. Soc. Amer. J.*, 11(3), 1964.

Devereux, R. F., J. W. Petre, and R. F. Kosic. "Real Time Oceanography: the Monster Buoy." *Mar. Sci. Instrum.*, 4, 1968.

Ellito, F. E. "Choosing Oceanographic Platforms." *Instrum. Soc. Amer. J.*, 11(3), 1964.

Grant, D. A. "Current, Temperature, and Depth Data from a Moored Oceanographic Buoy." *Can. J. Earth Sci.*, 5(5), 1968.

Inyutkina, A. I. "Electronic Buoys for Data Collection Concerned with the Labrador Current." (In Russian). *Okeanol., Akad. Nauk, SSSR*, 4(3), 1964.

Kielhorn, W. V., W. S. Richardson, and F. P. Burke. "Bathythermograph Measurements from Fixed-Wing Aircraft." *J. Mar. Techn. Soc.,* 5(2), 1971.

Nath, J. H., and S. Neshyba. "Two-Point Mooring System for Spar Buoy." *J. Wat. Ways Harb. Coast. Eng. Div. Am. Soc. Civ. Engrs,* 97(WW2), 1971.

Parker, E. K. "SCUBA as a Tool for Scientists." *Mar. Sci. Instrum.,* 1, 1962.

Petre, J. W., and R. F. Devereux. "The Monster Buoy—a Progress Report for 1967." *Mar. Sci. Instrum.,* 4, 1968.

Richardson, W. S., P. B. Stimson, and C. H. Wilkins. "Current Measurements from Moored Buoys." *Deep-Sea Res.,* 10(4), 1963.

Ruff, R. E. "Self-Contained Oceanographic Data Acquisition Buoy." *Mar. Sci. Instrum.,* 4, 1968.

Shonting, D. H., and A. H. Barrett. "A Stable Spar-Buoy Platform for Mounting Instrumentation." *J. Mar. Res.,* 29(2), 1971.

Stass, I. I. "Performance of Hydrographic Buoy Stations." (In Russian). *Trudy Morsk. Gidrofiz. Inst.,* 26, 1962. Transl.: Scripta Tecnica, 26, 1964.

Walden, R. C. "A Review of Oceanographic and Meteorological Buoy Capabilities and Effectiveness." *Mar. Sci. Instrum.,* 4, 1968.

Warhurst, J. S. "A Free-Diving Oceanographic Buoy." *Symposium, Ocean Sci. and Ocean Eng.* Mar. Techn. Soc., Am. Soc. Limnol. Oceanogr., 2, 1965.

Wyatt, F. G., Jr. "Buzzards Bay Satellite Spar Buoy for the Study of Air-Sea Interaction." *Mar. Sci. Instrum.,* 4, 1968.

CHAPTER 13

Data Transmission and Analysis

1301. Data Transmission

Once the data from various oceanographic sensors are obtained they must be transmitted to some central location for collection and analysis so that they can be used to some advantage. Whether this advantage is in the form of proving or disproving some esoteric mathematical model or in the form of some operational forecast is really of no importance. The primary concern is that the data be used.

1302. Cable Links

The most common method of transmitting data to at least the primary analysis or primary collection stage is by electrical cable; and the type of cable most commonly used is the multi-conductor cable, which requires a relatively simple system because, since each sensor has its own channel, there is no need to separate the channels. However, this saving in complexity is often more than offset by the amount of room this large-diameter cable occupies aboard a ship where space is at a premium. Thus, the depths to which many instruments can be lowered are limited by cable requirements.

Storage aboard ship is not the only problem with cable. When a large-diameter cable is payed out, it presents to moving water a large cross-sectional area and therefore generates a certain amount of drag, making it difficult, if not impossible, to lower a device to any great depth. The cable just streams out behind the vessel and, unless a depth-measuring sensor is available, the true instrument depth is greatly in doubt.

A single-conductor cable can be very small in diameter and, therefore, present much less drag to moving water. However, when such a cable is used, the complexity of the system increases because there have to be circuits to separate the various signals and the source power. Single-conductor cables are usually used in connection with **FM** systems wherein each individual sensor sends its data up the cable as a varying frequency that is separated out at the top and converted into digital form by proper circuitry.

In any event, the cable is the most commonly used data link, primarily because it is the most dependable. Although leakage and breakage problems both occur with some frequency, the reliability of a cable data link is still unquestionably greater than that of any other data link available.

1303. Acoustic Links

Another method of transmitting data is by an acoustic link, that is, by passing sound energy through the water. This method has not been employed extensively, probably because it is affected by environmental factors. In most cases, either its range is limited because of such conditions as varying temperature structures, or its signal-to-noise ratio is somewhat less than optimum because of ambient noise. Furthermore, since attenuation losses increase with increasing frequency, data transmission is limited to relatively low frequencies. Very simply, this means that the number of data bits per unit time that can be transmitted is relatively small because the bandwidth available at these low frequencies is small.

One way of getting around the bandwidth limitation is to extend the data-sending time even when the data have been acquired within a short time period. For example, a television image, which inherently requires a great deal of data, can be transmitted acoustically by means of storage techniques and a slow-scan type of image. Of course, if the object being photographed were moving during this scan the resulting image would be unsatisfactory.

An additional shortcoming is caused by the short-term changes in the environment that often occur and decrease the dependability of acoustic links to a greater extent than is the case with some of the other methods of data transmission. The acoustic links appear to be satisfactory for ranges on the order of a few hundred yards, but for longer ranges their dependability and practicability for data transmission are somewhat dubious.

1304. Radio Links

Radio frequencies are also commonly used for transmitting data. The major advantage of radio-link transmission is the fact that it is relatively high in frequency and therefore has a high bandwidth capability, which allows large numbers of data bits to be transmitted. Although radio links are very commonly used in conjunction with buoys, there are two major drawbacks to this mode of telemetry: one has to do with varying transmission capability produced by atmospheric electrical activity, while the other concerns antenna attitude. As a buoy bounces around on the sea surface, its radio antenna sometimes points in directions not conducive to ideal radio transmission, resulting in periods when transmission is not only poor but intermittent. For these reasons, provision is usually made either to transmit data more than once or to continue transmission until it is recognized by the receiver.

Probably the most dependable type of radio link is one that uses a satellite either as a collector that telemeters previously stored data to a collection station, or as a relay station for a buoy's radio link in much the same manner as communications satellites operate.

1305. Laser Links

Another way in which data might possibly be transmitted is by laser energy, but there are disadvantages. A laser is strictly a line-of-sight device and its transmission range is therefore limited. Also, because a laser is fixed in frequency, the only type of information vehicle available is amplitude modulation. In addition, light does not penetrate clouds or rain as effectively as do other forms of electromagnetic radiation, such as radio waves, and, at this time, lasers have high power requirements because of their relative inefficiency. For all these reasons, it does not appear as if lasers will be used extensively for the transmission of oceanographic data within the foreseeable future.

1306. Analysis Considerations

Once data have arrived at a collection center, the question of analysis arises. Analysis problems can best be solved by designing data inputs so that analysis may be done in the most direct manner, preferably by automatic methods. Thus, if the sensor system is designed to produce an output compatible with a computer input, analysis is comparatively simple. Modern devices are so designed, and many of them immediately condition the signal so that the readout appears on computer-compatible magnetic or punched-paper tape. In this form, computer analysis can be performed aboard ship, if facilities are available, or with minimum expenditure of time and effort when the tape is brought ashore.

However, some forms of data do not readily lend themselves to computer analysis: identification of plankton forms and classification of bottom sediments are two examples of these types of data. Most physical data, however, are suitable for computer analysis, and more and more modern vessels are being outfitted with self-contained computer systems. Many of the newer oceanographic vessels, for example, have computer systems that accept magnetic and gravity data along with a navigational input, so that a real time plot of the magnetic and gravity fields is generated as the vessel proceeds through its cruise schedule.

1307. Reliability

It may be seen that the rather conspicuous trend toward complexity in oceanographic instrumentation results from attempts to acquire as complete a picture of the environment as possible, and to make that picture as accurate as possible.

However, the experience that many oceanographers have had in taking complex equipment to sea has not been a very pleasant one, and so there is also a trend toward simplicity. This trend, of course, is motivated by two desires: reliability and low cost.

Electronic systems will become more and more reliable and, when they reach the point where their failure rate at sea is no worse than that of such tried and true non-electronic devices as the Nansen bottle and the Secchi disc, the oceanographer's aversion to them will be overcome.

1308. Look Out the Window

As gear becomes more and more sophisticated and automatic, there is a natural tendency to assume that the machine is smarter than its operator. In some cases this is true, but usually it is not. The operator has to exercise even more care to insure that errors do not multiply than he does when the instruments are simple and easy to understand.

Both the designer and the user of environmental instruments should constantly be reminded of the existence of an "instrument" capability which, although not quite as elegant or esoteric as an STD (Salinity, Temperature, Depth), still serves as an excellent backup. This capability, consisting of our five senses, might be termed the "Look Out the Window Instrumentation Suite." Although not capable of $\pm 1\%$ accuracy, the eyes can determine approximate wind speed from the configuration of the sea surface. The nose can often detect the presence of hydrogen sulphide. The taste can differentiate between salt and fresh water. The index finger, properly deployed, can discern warm and cool water. The ears can estimate wave period by listening for water-slap on a ship's hull.

This is not to imply that mechanical, chemical, and electronic devices should be supplanted by the "Look Out the Window Instrumentation Suite." What is suggested is that the senses form a necessary and inexpensive complement to other pieces of gear. Many a set of weird data would have been nipped in the bud if the investigator had only bothered to "look out the window."

1309. State of the Art

It is becoming more and more apparent that the day of the oceanographic vessel that not only collects but also analyzes data in real time is very close at hand. It should not be too long before a vessel will go to sea, acquire data, and, while still at sea, completely reduce these data for *all* the parameters discussed in this book. This will be in marked contrast to the situation existing today at some laboratories, where it takes upwards of a year to analyze completely the results of a cruise.

In this manner it is hoped that as the data accumulated about the oceans increase, understanding of them will at least increase at the same rate.

Sources and Additional Reading

Abbott, J. L., G. S. Morris, Jr., and J. D. Mudie. "Scripps' Seagoing Computer Centers." *Trans. Applications of Sea Going Computers Symposium,* Mar. Tech. Soc., 1969.

Adams, R. L., and R. R. Gardner. "Computer Processing of Acoustic Data at Sea." *J. Acoust. Soc. Am.,* 39(4), 1966.

Branham, D. W. "Shipboard Oceanographic Survey System." *Symposium, Ocean Sci. and Ocean Eng.* Mar. Techn. Soc., Am. Soc. Limnol. Oceanogr., 1, 1965.

———. "The Shipboard Computer's Place in Oceanography." *Mar. Sci. Instrum.,* 4, 1968.

Canney, H., Jr. "A Buoy-Satellite System for Current Measurement." *Undersea Techn.,* 5(3), 1964.

Crease, J. "Experiences with a Computer in Oceanographic Research at Sea." *J. Inst. Navig. Lond.,* 24(3), 1971.

Ewing, G. C. "On the Design Efficiency of Rapid Oceanographic Data-Acquisition Systems." *Deep-Sea Res.,* Suppl. 16, 1969.

Furuhata, T. "New Automatic Data Processing Machine, MERIAC-1-F." *J. Oceanogr. Soc.,* Japan, 18(3), 1962.

Howe, M. R., and R. I. Tait. "An Evaluation of an In-Situ Salinity-Temperature-Depth Measuring System." *Mar. Geol.,* 3(6), 1965.

Jones, E. E. "Computer Data-Acquisition Systems Aboard ESSA Survey Ships." Trans. *Applications of Sea Going Computers Symposium,* Mar. Tech. Soc., 1969.

Kamenskaya, O. A. "About the Use of Electronic Computers in Oceanological Computations." (In Russian). *Mat. Vtoroi Konf., Vzaimod. Atmos. Gidrosfer. v. Severn. Atlant. Okean., Mezhd. Geofiz. God.,* Leningr. Gidromet. Inst., 1964.

Kanari, S., "Underwater Acoustic Telemetry for Oceanographical and Limnological Research." Parts 1 and 2. *Bull. Disaster Prevention Res. Inst.,* Kyoto Univ., 15(3), 1966.

Ketchum, D. D., and R. G. Stevens. "A Data Acquisition and Reduction System for Oceanographic Measurements." *Mar. Sci. Instrum.,* 1, 1962.

Long, F. S. "Redundancy Gives Researcher Computer System High Reliability." *UnderSea Technol.,* 10(8), 1969.

McLoon, C. "The Problems of Reliable Long-Range Transmission of Remote Oceanographic Measurements." *Mar. Sci. Instrum.,* 1, 1962.

Mitson, R. B., P. G. Griffiths, and C. R. Hood. "An Underwater Acoustic Link." *Deep-Sea Res.,* 14(2), 1967.

Murray, M. T. "Tidal Analysis with an Electronic Computer." *Cah. Océanogr.,* CCOEC, 15(10), 1963.

O'Hagan, R. M., and P. F. Smith. "The Use of a Small Computer with Real Time Techniques for Oceanographic Data Acquisition, Immediate Analysis and Presentation." *Symposium, Ocean Sci. and Ocean Eng.* Mar. Techn. Soc., Am. Soc. Limnol. Oceanogr., 1, 1965.

Paramonov, A. N., G. G. Neuymin, V. I. Man'kovskiy, and W. A. Prokhorenko. "Hydroacoustic System for Telemetering Transparency of sea water." (In Russian). In: *Metody i pribory dlya Issledovaniya Fizicheskikh Protsessov v Okeane*, A. N. Paramonov, ed., Izd-vo Naukova Dumka, Kiev, 1966. Transl.: JPRS: 39, 13 Feb. 1967 (Clearinghouse Fed. Sci. Tech. Info., U.S. Dept. Commerce).

Pederson, A. M. "Spurv Instrumentation System." *Mar. Sci. Instrum.*, 4, 1968.

Pingree, R. D. "Regularly Spaced Instrumented Temperature and Salinity Structures." *Deep-Sea Res.*, 18, 1971.

Rogozin, A. A. "The Experience of Use of the Electronic Calculator for the Determination of Elements of the Waves by the Stereophotogrammetrical Method." (In Russian). *Met. i. Gidrol.*, 9, 1963.

Shekhtman, A. N. "On the Machine Treatment of Oceanographic Data." (In Russian; English abstract). *Okeanologiia*, 9(3), 1969.

Shipek, Carl J. "A New Deep-Sea Oceanographic System." *Symposium, Ocean Sci. and Ocean Eng.* Mar. Techn. Soc., Am. Soc. Limnol. Oceanogr., 2, 1965.

Snodgrass, F. E. "Deep-Sea Instrument Capsule." *Science*, 162(3849), 1968.

Snodgrass, J. M. "Long-Range Outlook for Oceanographic Telemetering." *Mar. Sci. Instrum.*, 1, 1962.

Stevenson, R. E., R. D. Terry, and J. S. Bailey. "The Photography from Space of Oceanographic Features." *Mar. Sci. Instrum.*, 4, 1968.

Stoertz, C. R. "Airborne Oceanography." *J. Mar. Techn. Soc.*, 3(2), 1969.

Talwani, M., J. Dorman, and R. Kittredge. "Experiences with Computers Aboard Research Vessels *Vema* and *Robert D. Conrad*." *Trans. Applications of Sea Going Computers Symposium*, Mar. Tech. Soc., 1969.

Walden, R. C., and D. H. Frantz. "A Long-Range, Oceanographic Telemetering System." *Mar. Sci. Instrum.*, 1, 1962.

Williams, J. "A Simplified Approach to Instrumentation." *Proc. 7th Annual Conf. Marine Technology Society*, 1971.

Wylde, J. A. "Transmitting a Number of Sensor Signals Through a Single Underwater Cable." *Mar. Sci. Instrum.*, 3, 1965.

APPENDIX A

Electricity Review

A-1. Charge and Current

The basic quantity involved in electricity is the *coulomb,* a measure of charge. The charge contained in a coulomb is equivalent to that of 6.25×10^{18} electrons. When charge is caused to move, a current is said to exist, and a flow of charge of one coulomb per second past any point is equivalent to a current of one ampere. This current flow may be in a constant direction, direct current (DC); or it may be in alternating directions, alternating current (AC). Direct current is analogous to the flow of a river continually moving water in one direction, whereas alternating current is analogous to the flow in a tidal estuary wherein the water is moved first upstream and then downstream.

A-2. Potential

In order to produce a flow of current, a source of energy is required. This source of energy is called *electrical potential* and is defined as the amount of energy per unit charge measured in volts, where one volt is equivalent to one joule per coulomb. Thus, electrical potential, voltage, and electromagnetic force (EMF) are ways of expressing the potential of a circuit for doing work.

A-3. Ohm's Law

When a difference in potential exists and there is a tendency for current to flow from one point in an electrical circuit to another, this current may be hindered by an electrical quantity called *resistance*. Resistance is expressed in ohms and is related to potential and current by Ohm's law

$$E = IR \tag{1}$$

where E = electrical potential in volts
I = current in amperes
R = resistance in ohms.

A-4. Power

Whenever there is a current flow through a resistance, some of the electrical energy is converted into heat. This heat conversion may be described in terms of the power dissipated in the resistance

$$P = I^2R \qquad (2)$$

where P = power in watts.

In a circuit that not only dissipates heat but does mechanical work, the power used can be determined from the electrical quantities by simply substituting the proper form of Equation (1) into Equation (2). Thus power may also be described as

$$P = EI \qquad (3)$$

or
$$P = E^2/R \qquad (4)$$

A-5. Capacitance

If two metallic plates are separated by some non-conducting material and an EMF is imposed across them, a charge is built up on one of the plates and this configuration is called a *capacitor*. The amount of charge accumulated on the plates of a capacitor with a given potential difference is a function of the area of the plates, their spacing, and the dielectric constant of the material separating them. The capacitance of a capacitor can also be related to the charge built up and the potential difference as follows

$$C = \frac{Q}{E} \qquad (5)$$

where C = capacitance in farads
Q = charge in coulombs
E = potential difference across the plates in volts.

A-6. Inductance

When charge is caused to flow through a wire, a compass placed near the wire deflects, indicating the presence of a magnetic field and the basic relationship between electricity and magnetism. Wherever there is a moving charge, there is an associated magnetic field. Conversely, a changing magnetic field always produces an electric field or a tendency to move charge (current). If the wire through which the charge is flowing is formed into a coil, the magnetic field associated with the current in the wire is concentrated in a smaller volume, so that when the current is first turned on there is a tendency to offset the generation of the steady-state magnetic field. This tendency toward the suppression of a changing magnetic field is simply another case of energy being stored. The amount of energy that can be stored within a coil, or the extent to which the coil tends to prevent a change in current through it, is a measure of its inductance. *Inductance* is therefore defined as the stored energy per unit time change in current, or mathematically

$$L = \frac{E}{dI/dt} \tag{6}$$

where L = inductance of the coil in henrys
E = stored energy per unit charge in volts
dI/dt = rate of change of current with respect to time in amperes per second.

A coil's inductance can be easily determined in another manner since it is related to the number of turns within a given area, the size of the coil, and the material about which the wire is wound.

A-7. Resistances in Series

There are two basic configurations by which electrical components can be connected: series connections and parallel connections. Series connections are so oriented that the current flowing through one component will flow through all the components in the series configuration, as is shown in Figure A-1. The total resistance in a group of resistances in series is determined by simply adding the components. Thus, total resistance between points "a" and "b" in Figure A-1 (R_{st}) is obtained by

$$R_{st} = R_1 + R_2 + R_3 \tag{7}$$

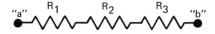

Figure A-1. Three resistors in series.

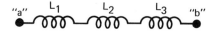

Figure A-2. Three inductors in series.

A-8. Inductances in Series

Inductances in series are treated in much the same manner as resistances, that is, the total inductance between points "a" and "b" (L_{st}) in the circuit in Figure A-2, is obtained by

$$L_{st} = L_1 + L_2 + L_3 \tag{8}$$

A-9. Capacitances in Series

Capacitors, on the other hand, act somewhat differently. The total capacitance of a series capacitance network is not the simple sum of the individual capacitances. In order to obtain the total capacitance between points "a" and "b" (C_{st}) in Figure A-3, the reciprocals must be added

$$\frac{1}{C_{st}} = \frac{1}{C_1} + \frac{1}{C_2} + \frac{1}{C_3} \tag{9}$$

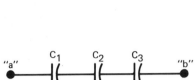

Figure A-3. Three capacitors in series.

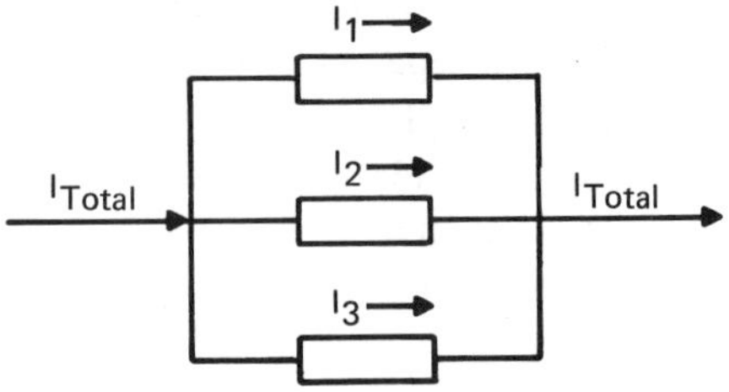

Figure A-4. Current divided in a parallel net.

A-10. Resistances in Parallel

In contrast to the series circuit, where the current that passes through one component passes through all the others, in a parallel circuit the current is split. Figure A-4 shows three resistors in parallel. The amount of current flowing through any one of the resistors is a function of the size of the resistor and has no relation to the amount of current flowing through any of the other resistors in the system. The total resistance of a parallel net is determined by Equation (10) where the resistance between the points "a" and "b" (R_{pt}) is now determined (*see* Figure A-5), using a relationship similar to that for capacitors in series

$$\frac{1}{R_{pt}} = \frac{1}{R_1} + \frac{1}{R_2} + \frac{1}{R_3} \qquad (10)$$

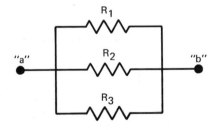
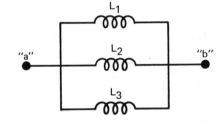

Figure A-5. Three resistances in parallel.

Figure A-6. Three inductors in parallel.

A-11. Inductances in Parallel

Inductances in parallel are treated in much the same manner as resistances in parallel (*see* Figure A-6), as is shown by Equation (11), which relates the total inductance of a number of parallel inductances (L_{pt})

$$\frac{1}{L_{pt}} = \frac{1}{L_1} + \frac{1}{L_2} + \frac{1}{L_3} \qquad (11)$$

A-12. Capacitances in Parallel

Placing capacitors in parallel is essentially the same thing as adding the area of the plates of the individual capacitors, so that parallel capacitances are simply added to obtain the total capacitance of the network (*see* Figure A-7). In other words, the capacitance of a number of capacitors in parallel (C_{pt}) is given by

$$C_{pt} = C_1 + C_2 + C_3 \qquad (12)$$

A-13. Resonant Circuits

A capacitance and an inductance combined in one circuit form a resonant circuit and, for various reasons, the frequency of this resonant circuit is given by

$$f = \frac{1}{2\pi\sqrt{LC}} \qquad (13)$$

where f = resonant frequency in hertz
L = inductance in henrys.
C = capacitance in farads.

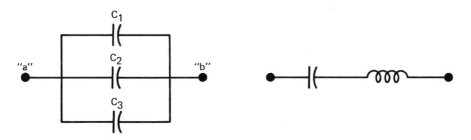

Figure A-7. Three capacitors in parallel.   Figure A-8. A series resonant circuit.

Equation (13) describes both a series and a parallel resonant circuit. These two resonant circuits have different characteristics, especially with respect to their impedance at resonance. At resonance, a series circuit (*see* Figure A-8) has a minimum value of impedance, while a parallel resonant circuit (*see* Figure A-9) has a maximum value of impedance. At the resonant frequency determined by Equation (13), a series resonant circuit will therefore exhibit a large current value, while a parallel resonant circuit will have a large voltage across it.

A-14. Voltage Measurement

The device usually used for direct measurement of current and voltage is a galvanometer of one sort or another. Galvanometers are actually sensitive to DC currents but, by use of proper circuitry, can be used for AC current and DC and AC voltage as well. However, if more accuracy is desired, other methods may be employed to measure voltage or current rather than the direct-reading meter movement.

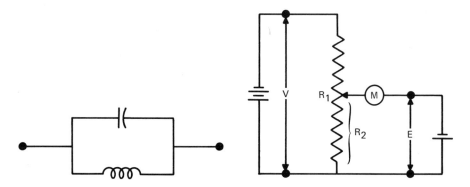

Figure A-9. A parallel resonant circuit.   **Figure A-10.** The potentiometer method of measuring voltage.

One of these instruments is the potentiometer, a simple circuit designed to measure voltage (*see* Figure A-10). In actuality, a potentiometer does not measure voltage: it compares an unknown voltage with a known one. When the moving arm of the potentiometer reaches a point where the current flow through the meter is zero, then

$$\frac{V}{E} = \frac{R_1}{R_2} \qquad (14)$$

The variable resistance R_2 is usually calibrated in percentages, so that the ratio of R_1 to R_2 is read directly from the dial. In this manner, the ratio is simply multiplied by the value of the known voltage to obtain the unknown voltage. The advantage of this circuit is that, since at null there is no current flow through the system, the reference battery lasts for a long time.

A-15. Resistance Measurements

Another parameter very often measured in instrumentation is electrical resistance. This can be done by simply using a known voltage to measure the current flowing through the unknown resistance, as in the ohmmeter shown in Figure A-11.

A somewhat more elegant way of measuring resistance is with a bridge circuit typified by the Wheatstone bridge shown in Figure A-12. At null, the Wheatstone bridge equation is

$$\frac{R_1}{R_2} = \frac{R_x}{R_3} \qquad (15)$$

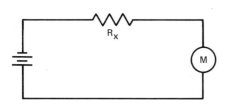 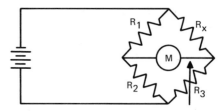

Figure A-11. The Ohmmeter method of measuring resistance.

Figure A-12. The Wheatstone bridge method of measuring resistance.

As may be seen from Equation (15), if R_1 is made equal to R_2, then the unknown resistance is simply equal to the value of the variable resistance as adjusted for null.

Another method of using the bridge is to adjust it for null at one end of the excursion of the unknown resistance and allow the imbalance to be read on the meter; in this way, it is not necessary to readjust for null each time. This is an especially useful way of measuring resistance when the variation of resistance to be measured is non-linear. Since the output of the bridge is also non-linear, the two non-linearities sometimes cancel each other, resulting in a linear meter-scale. This is often done when temperature is being measured with thermistors, in which case the temperature is indicated directly on the meter face.

Although both the potentiometer and the bridge are basically DC circuits, they are often used in AC circuits. Therefore, many rather sophisticated instruments have either a potentiometer or a bridge as their basic measuring circuit.

APPENDIX B

Chlorinity Determined from Temperature and Conductivity

The chlorinity in parts per thousand is determined from the product of χ and γ:

$$(\chi)\,(\gamma) \approx \text{Cl}^0/_{00}$$

Part I—Temperature in °C (θ) as a function of χ

θ	χ	θ	χ	θ	χ	θ	χ
−2.0	1.4870	2.0	1.2977	6.0	1.1493	10.0	1.0262
−1.9	1.4815	2.1	1.2935	6.1	1.1460	10.1	1.0233
−1.8	1.4760	2.2	1.2895	6.2	1.1426	10.2	1.0205
−1.7	1.4707	2.3	1.2854	6.3	1.1394	10.3	1.0177
−1.6	1.4653	2.4	1.2814	6.4	1.1360	10.4	1.0150
−1.5	1.4600	2.5	1.2774	6.5	1.1327	10.5	1.0123
−1.4	1.4547	2.6	1.2733	6.6	1.1294	10.6	1.0095
−1.3	1.4495	2.7	1.2694	6.7	1.1261	10.7	1.0068
−1.2	1.4444	2.8	1.2655	6.8	1.1230	10.8	1.0040
−1.1	1.4393	2.9	1.2616	6.9	1.1197	10.9	1.0013
−1.0	1.4341	3.0	1.2576	7.0	1.1165	11.0	.9986
−0.9	1.4291	3.1	1.2537	7.1	1.1132	11.1	.9959
−0.8	1.4241	3.2	1.2499	7.2	1.1100	11.2	.9932
−0.7	1.4191	3.3	1.2461	7.3	1.1069	11.3	.9906
−0.6	1.4143	3.4	1.2423	7.4	1.1038	11.4	.9879
−0.5	1.4094	3.5	1.2385	7.5	1.1006	11.5	.9852
−0.4	1.4045	3.6	1.2346	7.6	1.0974	11.6	.9826
−0.3	1.3997	3.7	1.2309	7.7	1.0943	11.7	.9800
−0.2	1.3949	3.8	1.2272	7.8	1.0913	11.8	.9774
−0.1	1.3903	3.9	1.2235	7.9	1.0882	11.9	.9747
0.0	1.3855	4.0	1.2198	8.0	1.0851	12.0	.9722
0.1	1.3808	4.1	1.2160	8.1	1.0820	12.1	.9696
0.2	1.3762	4.2	1.2124	8.2	1.0789	12.2	.9670
0.3	1.3716	4.3	1.2088	8.3	1.0759	12.3	.9644
0.4	1.3670	4.4	1.2052	8.4	1.0728	12.4	.9619
0.5	1.3624	4.5	1.2016	8.5	1.0699	12.5	.9593
0.6	1.3579	4.6	1.1979	8.6	1.0669	12.6	.9567
0.7	1.3534	4.7	1.1944	8.7	1.0639	12.7	.9542
0.8	1.3489	4.8	1.1908	8.8	1.0690	12.8	.9517
0.9	1.3446	4.9	1.1873	8.9	1.0579	12.9	.9492
1.0	1.3401	5.0	1.1838	9.0	1.0550	13.0	.9467
1.1	1.3357	5.1	1.1802	9.1	1.0520	13.1	.9443
1.2	1.3314	5.2	1.1767	9.2	1.0491	13.2	.9418
1.3	1.3271	5.3	1.1733	9.3	1.0462	13.3	.9393
1.4	1.3229	5.4	1.1698	9.4	1.0432	13.4	.9368
1.5	1.3186	5.5	1.1664	9.5	1.0404	13.5	.9344
1.6	1.3143	5.6	1.1629	9.6	1.0376	13.6	.9320
1.7	1.3102	5.7	1.1595	9.7	1.0347	13.7	.9296
1.8	1.3059	5.8	1.1561	9.8	1.0318	13.8	.9271
1.9	1.3018	5.9	1.1527	9.9	1.0290	13.9	.9247

CHLORINITY-CONDUCTIVITY TABLES

θ	χ	θ	χ	θ	χ	θ	χ
14.0	.9224	18.3	.8282	22.6	.7489	26.9	.6812
14.1	.9200	18.4	.8262	22.7	.7472	27.0	.6798
14.2	.9178	18.5	.8243	22.8	.7455	27.1	.6783
14.3	.9152	18.6	.8222	22.9	.7438	27.2	.6769
14.4	.9128	18.7	.8203	23.0	.7421	27.3	.6754
14.5	.9105	18.8	.8183	23.1	.7405	27.4	.6739
14.6	.9082	18.9	.8163	23.2	.7387	27.5	.6725
14.7	.9059	19.0	.8144	23.3	.7371	27.6	.6711
14.8	.9035	19.1	.8125	23.4	.7354	27.7	.6698
14.9	.9012	19.2	.8105	23.5	.7338	27.8	.6682
15.0	.8989	19.3	.8086	23.6	.7321	27.9	.6668
15.1	.8967	19.4	.8066	23.7	.7305	28.0	.6654
15.2	.8943	19.5	.8047	23.8	.7288	28.1	.6640
15.3	.8921	19.6	.8029	23.9	.7271	28.2	.6625
15.4	.8898	19.7	.8009	24.0	.7255	28.3	.6612
15.5	.8876	19.8	.7990	24.1	.7239	28.4	.6598
15.6	.8853	19.9	.7971	24.2	.7222	28.5	.6582
15.7	.8831	20.0	.7952	24.3	.7206	28.6	.6568
15.8	.8809	20.1	.7934	24.4	.7190	28.7	.6555
15.9	.8786	20.2	.7915	24.5	.7175	28.8	.6541
16.0	.8764	20.3	.7897	24.6	.7158	28.9	.6526
16.1	.8743	20.4	.7878	24.7	.7142	29.0	.6513
16.2	.8720	20.5	.7860	24.8	.7126	29.1	.6499
16.3	.8699	20.6	.7841	24.9	.7110	29.2	.6484
16.4	.8677	20.7	.7823	25.0	.7094	29.3	.6471
16.5	.8656	20.8	.7806	25.1	.7079	29.4	.6457
16.6	.8634	20.9	.7787	25.2	.7063	29.5	.6442
16.7	.8613	21.0	.7769	25.3	.7048	29.6	.6428
16.8	.8592	21.1	.7751	25.4	.7034	29.7	.6414
16.9	.8570	21.2	.7733	25.5	.7019	29.8	.6400
17.0	.8549	21.3	.7715	25.6	.7003	29.9	.6387
17.1	.8528	21.4	.7698	25.7	.6988	30.0	.6372
17.2	.8507	21.5	.7680	25.8	.6974	30.1	.6359
17.3	.8486	21.6	.7662	25.9	.6959	30.2	.6345
17.4	.8465	21.7	.7645	26.0	.6943	30.3	.6330
17.5	.8445	21.8	.7627	26.1	.6929	30.4	.6317
17.6	.8424	21.9	.7609	26.2	.6915	30.5	.6303
17.7	.8404	22.0	.7592	26.3	.6899	30.6	.6290
17.8	.8383	22.1	.7575	26.4	.6885	30.7	.6275
17.9	.8362	22.2	.7557	26.5	.6870	30.8	.6261
18.0	.8342	22.3	.7540	26.6	.6855	30.9	.6248
18.1	.8322	22.4	.7523	26.7	.6841	31.0	.6234
18.2	.8302	22.5	.7506	26.8	.6826		

Part II—Conductivity in millimhos per cm (G) as a function of γ

G	γ	G	γ	G	γ	G	γ
0.0	0.000	1.6	0.613	3.2	1.288	4.8	1.990
0.1	0.031	1.7	0.653	3.3	1.331	4.9	2.035
0.2	0.066	1.8	0.695	3.4	1.374	5.0	2.079
0.3	0.102	1.9	0.736	3.5	1.417	5.1	2.124
0.4	0.139	2.0	0.778	3.6	1.461	5.2	2.168
0.5	0.176	2.1	0.820	3.7	1.505	5.3	2.213
0.6	0.214	2.2	0.862	3.8	1.549	5.4	2.258
0.7	0.253	2.3	0.904	3.9	1.592	5.5	2.303
0.8	0.292	2.4	0.946	4.0	1.636	5.6	2.348
0.9	0.331	2.5	0.988	4.1	1.680	5.7	2.393
1.0	0.370	2.6	1.031	4.2	1.724	5.8	2.438
1.1	0.410	2.7	1.073	4.3	1.768	5.9	2.484
1.2	0.450	2.8	1.116	4.4	1.812	6.0	2.529
1.3	0.490	2.9	1.159	4.5	1.856	6.1	2.575
1.4	0.531	3.0	1.201	4.6	1.901	6.2	2.620
1.5	0.572	3.1	1.244	4.7	1.945	6.3	2.665

G	γ	G	γ	G	γ	G	γ
6.4	2.711	13.1	5.871	19.8	9.195	26.5	12.646
6.5	2.757	13.2	5.920	19.9	9.246	26.6	12.699
6.6	2.802	13.3	5.969	20.0	9.297	26.7	12.751
6.7	2.848	13.4	6.017	20.1	9.347	26.8	12.803
6.8	2.894	13.5	6.066	20.2	9.398	26.9	12.856
6.9	2.940	13.6	6.114	20.3	9.449	27.0	12.908
7.0	2.985	13.7	6.163	20.4	9.499	27.1	12.961
7.1	3.031	13.8	6.212	20.5	9.550	27.2	13.013
7.2	3.077	13.9	6.261	20.6	9.601	27.3	13.066
7.3	3.123	14.0	6.310	20.7	9.652	27.4	13.119
7.4	3.170	14.1	6.358	20.8	9.703	27.5	13.171
7.5	3.216	14.2	6.407	20.9	9.754	27.6	13.224
7.6	3.262	14.3	6.456	21.0	9.805	27.7	13.276
7.7	3.308	14.4	6.505	21.1	9.856	27.8	13.329
7.8	3.354	14.5	6.554	21.2	9.907	27.9	13.382
7.9	3.401	14.6	6.603	21.3	9.958	28.0	13.434
8.0	3.447	14.7	6.652	21.4	10.009	28.1	13.487
8.1	3.494	14.8	6.701	21.5	10.060	28.2	13.540
8.2	3.546	14.9	6.750	21.6	10.111	28.3	13.593
8.3	3.587	15.0	6.800	21.7	10.162	28.4	13.645
8.4	3.633	15.1	6.849	21.8	10.213	28.5	13.698
8.5	3.680	15.2	6.898	21.9	10.264	28.6	13.751
8.6	3.727	15.3	6.947	22.0	10.315	28.7	13.804
8.7	3.773	15.4	6.997	22.1	10.367	28.8	13.857
8.8	3.820	15.5	7.046	22.2	10.418	28.9	13.910
8.9	3.867	15.6	7.095	22.3	10.469	29.0	13.963
9.0	3.914	15.7	7.145	22.4	10.520	29.1	14.016
9.1	3.961	15.8	7.194	22.5	10.572	29.2	14.069
9.2	4.008	15.9	7.243	22.6	10.624	29.3	14.122
9.3	4.055	16.0	7.293	22.7	10.675	29.4	14.175
9.4	4.102	16.1	7.342	22.8	10.726	29.5	14.228
9.5	4.149	16.2	7.392	22.9	10.777	29.6	14.281
9.6	4.196	16.3	7.442	23.0	10.829	29.7	14.334
9.7	4.243	16.4	7.491	23.1	10.880	29.8	14.387
9.8	4.290	16.5	7.541	23.2	10.932	29.9	14.441
9.9	4.338	16.6	7.590	23.3	10.983	30.0	14.494
10.0	4.385	16.7	7.640	23.4	11.035	30.1	14.547
10.1	4.432	16.8	7.690	23.5	11.087	30.2	14.600
10.2	4.480	16.9	7.740	23.6	11.139	30.3	14.654
10.3	4.527	17.0	7.789	23.7	11.190	30.4	14.707
10.4	4.575	17.1	7.839	23.8	11.241	30.5	14.760
10.5	4.622	17.2	7.889	23.9	11.293	30.6	14.813
10.6	4.670	17.3	7.939	24.0	11.345	30.7	14.867
10.7	4.717	17.4	7.989	24.1	11.397	30.8	14.920
10.8	4.765	17.5	8.039	24.2	11.448	30.9	14.974
10.9	4.813	17.6	8.089	24.3	11.500	31.0	15.027
11.0	4.860	17.7	8.139	24.4	11.552	31.1	15.080
11.1	4.908	17.8	8.189	24.5	11.604	31.2	15.134
11.2	4.956	17.9	8.239	24.6	11.656	31.3	15.187
11.3	5.004	18.0	8.290	24.7	11.708	31.4	15.241
11.4	5.052	18.1	8.339	24.8	11.760	31.5	15.295
11.5	5.100	18.2	8.389	24.9	11.812	31.6	15.348
11.6	5.148	18.3	8.439	25.0	11.864	31.7	15.402
11.7	5.196	18.4	8.489	25.1	11.916	31.8	15.455
11.8	5.244	18.5	8.540	25.2	11.968	31.9	15.509
11.9	5.292	18.6	8.590	25.3	12.020	32.0	15.563
12.0	5.340	18.7	8.640	25.4	12.072	32.1	15.616
12.1	5.388	18.8	8.691	25.5	12.124	32.2	15.670
12.2	5.436	18.9	8.741	25.6	12.176	32.3	15.724
12.3	5.484	19.0	8.791	25.7	12.228	32.4	15.777
12.4	5.533	19.1	8.842	25.8	12.280	32.5	15.831
12.5	5.581	19.2	8.892	25.9	12.333	32.6	15.885
12.6	5.629	19.3	8.943	26.0	12.385	32.7	15.939
12.7	5.678	19.4	8.993	26.1	12.437	32.8	15.993
12.8	5.726	19.5	9.044	26.2	12.489	32.9	16.047
12.9	5.774	19.6	9.094	26.3	12.542	33.0	16.100
13.0	5.823	19.7	9.145	26.4	12.594	33.1	16.154

CHLORINITY-CONDUCTIVITY TABLES

G	γ	G	γ	G	γ	G	γ
33.2	16.208	39.9	19.873	46.6	23.634	53.3	27.491
33.3	16.262	40.0	19.928	46.7	23.690	53.4	27.549
33.4	16.316	40.1	19.984	46.8	23.756	53.5	27.607
33.5	16.370	40.2	20.039	46.9	23.812	53.6	27.665
33.6	16.424	40.3	20.095	47.0	23.868	53.7	27.724
33.7	16.478	40.4	20.150	47.1	23.924	53.8	27.782
33.8	16.532	40.5	20.206	47.2	23.980	53.9	27.840
33.9	16.586	40.6	20.261	47.3	24.036	54.0	27.899
34.0	16.641	40.7	20.317	47.4	24.092	54.1	27.957
34.1	16.695	40.8	20.372	47.5	24.148	54.2	28.016
34.2	16.749	40.9	20.428	47.6	24.204	54.3	28.074
34.3	16.803	41.0	20.484	47.7	24.261	54.4	28.132
34.4	16.857	41.1	20.539	47.8	24.318	54.5	28.191
34.5	16.912	41.2	20.595	47.9	24.375	54.6	28.250
34.6	16.966	41.3	20.651	48.0	24.432	54.7	28.308
34.7	17.020	41.4	20.707	48.1	24.490	54.8	28.367
34.8	17.074	41.5	20.762	48.2	24.547	54.9	28.425
34.9	17.129	41.6	20.818	48.3	24.604	55.0	28.484
35.0	17.183	41.7	20.874	48.4	24.661	55.1	28.542
35.1	17.237	41.8	20.930	48.5	24.719	55.2	28.601
35.2	17.292	41.9	20.986	48.6	24.776	55.3	28.660
35.3	17.346	42.0	21.041	48.7	24.833	55.4	28.718
35.4	17.401	42.1	21.097	48.8	24.890	55.5	28.777
35.5	17.455	42.2	21.153	48.9	24.948	55.6	28.836
35.6	17.509	42.3	21.209	49.0	25.005	55.7	28.894
35.7	17.564	42.4	21.265	49.1	25.062	55.8	28.953
35.8	17.619	42.5	21.321	49.2	25.120	55.9	29.012
35.9	17.673	42.6	21.377	49.3	25.177	56.0	29.071
36.0	17.728	42.7	21.433	49.4	25.235	56.1	29.129
36.1	17.782	42.8	21.489	49.5	25.292	56.2	29.188
36.2	17.837	42.9	21.545	49.6	25.350	56.3	29.247
36.3	17.891	43.0	21.601	49.7	25.407	56.4	29.306
36.4	17.946	43.1	21.658	49.8	25.465	56.5	29.365
36.5	18.001	43.2	21.714	49.9	25.522	56.6	29.424
36.6	18.055	43.3	21.770	50.0	25.580	56.7	29.483
36.7	18.110	43.4	21.826	50.1	25.637	56.8	29.542
36.8	18.165	43.5	21.882	50.2	25.695	56.9	29.601
36.9	18.225	43.6	21.938	50.3	25.752	57.0	29.660
37.0	18.274	43.7	21.995	50.4	25.810	57.1	29.719
37.1	18.329	43.8	22.051	50.5	25.868	57.2	29.778
37.2	18.384	43.9	22.107	50.6	25.925	57.3	29.837
37.3	18.439	44.0	22.163	50.7	25.983	57.4	29.896
37.4	18.494	44.1	22.220	50.8	26.041	57.5	29.956
37.5	18.549	44.2	22.276	50.9	26.099	57.6	30.014
37.6	18.604	44.3	22.332	51.0	26.156	57.7	30.079
37.7	18.659	44.4	22.389	51.1	26.214	57.8	30.132
37.8	18.714	44.5	22.445	51.2	26.272	57.9	30.192
37.9	18.769	44.6	22.502	51.3	26.330	58.0	30.251
38.0	18.824	44.7	22.558	51.4	26.388	58.1	30.310
38.1	18.879	44.8	22.614	51.5	26.445	58.2	30.369
38.2	18.934	44.9	22.671	51.6	26.503	58.3	30.429
38.3	18.989	45.0	22.728	51.7	26.561	58.4	30.488
38.4	19.044	45.1	22.784	51.8	26.619	58.5	30.547
38.5	19.099	45.2	22.841	51.9	26.677	58.6	30.606
38.6	19.154	45.3	22.897	52.0	26.735	58.7	30.666
38.7	19.209	45.4	22.954	52.1	26.793	58.8	30.725
38.8	19.264	45.5	23.010	52.2	26.851	58.9	30.785
38.9	19.320	45.6	23.067	52.3	26.909	59.0	30.844
39.0	19.375	45.7	23.124	52.4	26.967	59.1	30.903
39.1	19.430	45.8	23.180	52.5	27.025	59.2	30.963
39.2	19.485	45.9	23.237	52.6	27.083	59.3	31.022
39.3	19.541	46.0	23.294	52.7	27.141	59.4	31.082
39.4	19.596	46.1	23.351	52.8	27.200	59.5	31.144
39.5	19.651	46.2	23.408	52.9	27.258	59.6	31.201
39.6	19.707	46.3	23.465	53.0	27.316	59.7	31.260
39.7	19.762	46.4	23.522	53.1	27.374	59.8	31.320
39.8	19.817	46.5	23.578	53.2	27.432	59.9	31.380
						60.0	31.440

Index

Absorption
  light, 83
  sound, 105
Absorption coefficient
  light, 84
  sound, 106
  water, 85
AC, 171
Accelerometers, wave, 140
Acceptance angle, 87
Accuracy
  definition, 13, 14
  depth determination, 31
  gravity-anomaly determination, 146
  instrument, 7
  magnetic-anomaly determination, 148
  oxygen determination, 126
  pH determination, 128
  salinity requirements, 48
  salinometers, 50
  sound-speed determination, 120
  temperature requirements, 39
  wave-height determination, 141, 142
Acoustic data links, 166
Adiabatic heating, 38
ADP, 115
Aircraft, as instrument platforms, 159
Airy wave 131–3
Alpha meters, 95
Alternating current, 171
Ampere, definition, 171
Analog output, 72
Analyzers, 6
Aneroid elements, 28

Anomalous behavior, 14
Anomalies
  gravity, 145
  magnetic, 147
Anti-roll tanks, 156
Argus Island, 152
Arrested rotors, 73
Attenuation, light, 84, 93
Attenuation coefficient, light, 84
Average, 4
Average deviation, 11

Bandwidth, optical system, 88
Bar, definition, 27
Bathythermographs, 3, 28, 40
  expendable, 26, 41
Bathythermograph traces, 137
Beam transmittance, 86, 93
Beam transmittance meters, 88, 95–8
Bias, 14
Biological noises, 113
Bioluminescence, 101
Black body radiation, 43
Bourdon tubes, 28
BT, 40
Bubbles, sound scattering, 107
Buoys
  instrumentation package, 154–5
  moorings, 154
  SPAR, 153
  stable, 153
  surface-following, 153

Cable
  data links, 165
  length as measure of depth, 25–6
Calendar's formula, 34

Calibration
  elimination of systematic errors, 14
  transducer, 116
Capacitance, definition, 172
Captured drag, 73
Catamarans, as instrument platforms, 156
Centrifugal force, 61
Ceramics, as sonic transducers, 115
Channels, sound, 110
Characteristic length, 84
Charge, electronic, 171
Chemical measurements, use of, 123
Chlorinity, 47
  chlorinity-conductivity tables, 179
Clarke photometer, 94
Coded discs, 30
Coded output, 72
Compliant moorings, 154
Computer compatability, 167
Conductivity, 48, 50
  pressure variation, 52
  chlorinity-conductivity tables, 179
Conductivity sensors, 6, 51–3
Confidence limits, 12
Contouring, three-dimensional, 6
Convergence, 64
Coriolis force, 61
Corrosion, 4
Cosine collector, 87
Coulomb, definition, 171
Critical rays, 110
Current
  causes, 61
  direct determination, 68–72
  direction, 78
  drag, 74–5
  equations, 61–3
  indirect determination, 66–8
  magnitude, 63
  measuring practices, 66
Current, electrical, definition, 171
  measurement, 175
Current meters
  dynamic, 68
  static, 73–5
Cylindrical spreading, 108

Damping discs, 140

Data analysis, 167
Data display, 6
Data storage, 155
Data transmission links
  acoustic, 166
  cable, 165–6
  laser, 167
  radio, 166–7
Data use, 5
DC, 171
Decibar, definition, 27
Deep layer, 36
Deep-water waves, 132
Density, 27, 48
Density gradient, 67
Depth determination, 25–31
  by cable length, 25
  by fall rate, 26
  by pressure, 26–9
  by satellite, 25
  by sonic methods, 25
  by temperature difference, 27
Digital output, 72
Direct current, 171
Directional wave spectra, 135
Dissolved oxygen
  distribution, 123
  measurement, 125
Doppler effect, 76, 162
Drag coefficient, 74
Drift devices, 68
Ducts, sound, 110
Dynamite, as sound source, 114

Eddies, 64
Ekman current meter, 70
Ekman spiral, 62
Electrical circuit analogy, 16–7
Electrodeless conductivity sensors, 51–3
Electromagnetic force, 171
Electromagnetic sensor, 75
Electrostrictive materials, 115
EMF, 171
Emissivity, 43
Equation of motion, 61
Error
  definition, 13
  illegitimate, 13, 14

in current measurement, 63, 66
probable, 11
random, 13, 14
standard, 11
systematic, 13, 14
Eulerian direct methods of measuring current, 69–72
Expendable STDs, 54, 55
Extinction, 93

Fall rate in depth determination, 26
Farad, definition, 172
Feedback, 6
Fetch, 133
Filters, color, 91
Fish bites, 4
Fixed instrument platforms, 151, 159
Fixed moorings, 154
FLIP, 153
Flux gate magnetometers, 148
FM data transmission, 165
Fouling, 4
Freedom, degrees of, 10
Free-floating buoys, 154
Fresnel's reflectance equation, 91
Friction, effect on currents, 61–2
Fully arisen sea, 134
Fully developed sea, 134

Gal, definition, 146
Galvanometers, 175
Gas exploders, 114
Gaussian distribution, 10
GEK, 67
Geomagnetic electrokinetograph, 67
Gershun tube, 87–8
Geostrophic flow, 62
Glass pH electrodes, 127
Gradient flow, 62
Gratings, for selecting wavelength, 94
Gravity anomaly, 145
  measurement, 146
Gray body, 43
Gyres, 64

Helicopters, as instrument platforms, 159
Henry, definition, 173

Holography, acoustic, 111
Hunting, 78
Hydrogen pH electrodes, 127

Ice measurement, 101
Illegitimate error, 13, 14
Immersion effect, 91
Inductance, 172
Instruments
  automated, 6
  automatic, 6
  characteristics, 6
  definition, 3, 5
  depth measurement, 25–31
  design criteria, 3
  development, 5
  fluid-motion measurements, 61–79
  gravity measurements, 145–6
  light measurements, 83–102
  magnetic-field measurements, 146–8
  oxygen measurements, 123–6
  pH measurements, 126–8
  salinity measurements, 47–57
  sound measurements, 105–20
  temperature measurements, 33–43
  tide measurements, 141
  wave measurements, 131–41
Instrument systems, 5
Integrated light values, 101
Intensity, 107
Interfacial waves, 136
Internal waves, 136
International temperature scale, 34
Irradiance
  definition, 86
  measurement, 94–5
  relative, 92–3

Kirchoff's law, 16
Knot, definition, 78
Knudsen's curves, 112

Lagrangian methods of measuring currents, 68–9
Lambert's law, 83
Lasers, 98
  as data links, 167
  for ice measurement, 102
  for water-level measurement, 101

Light
    as a tool, 101–2
    absorption, 83–5
    bandwidth, 88–91
    instruments, 94–101
    losses, 83, 92–4
    reflectance, 91–2
    scattering, 85–6
Light paths, coded-disc, 30
Linear differential transformers, 30
Link, system, 6
Look Out the Window Instrumentation, 168

MAD gear, 148
Magnesium sulphate, 105
Magnetic anomaly, 146–7
    measurement, 148
Magnetic field, earth, 146
Magnetic tape, 72
Magnetostrictive materials, 115
Mean, definition, 10
Measurands, 5
Mixed layer, 36
Moorings, buoy, 154
Motion
    determination of, 66–79
    Ekman spiral, 62
    equation of, 61–2
    forces producing, 61
    geostrophic, 62
    gradient, 62
    magnitude, 63
    surface currents, 64–6
    tidal, 63
    vertical, 64
Moveable instrument platforms, 151, 155–9
Multi-conductor cables, 165

Nansen bottles, 27
National Data Buoy Project, 162
Naviface, 107
Navigation by satellite, 69
Nephelometers, 100
Noise
    ambient, 112–3
    self, 118

NR-1, 157
Nuclear precession magnetometers, 148

Ocean
    ambient noise, 112–3
    currents, 64
    homogeneity, 3
    oxygen, 123–5
    pH, 126–7
    salinity, 49
    size, 3
    temperature, 35–8
Ohm, 171
Ohm's law, 171
One-point moorings, 155
Oscillators, variable-frequency, 30
Overheated wire, 76
Oxygen
    distribution, 123–5
    electrodes, 125–6

Parallel circuits
    resistance, 174
    inductance, 174
    capacitance, 175
Pendulum, gravity, 146
Period, eddy, 64
    wave, 133
Permanent currents, 64
pH
    distribution, 126–7
    measurement, 127–8
pH electrodes
    glass, 127
    hydrogen, 127
    quinhydrone, 127
Photocells
    as sensors, 88, 94, 95
    as signal conditioners, 30
Photometer, Clarke, 94
Piezoelectric materials, 115
Pitot tubes, 73
Planck radiation, 34
Plankton determinations, 101
Platforms, instrument
    fixed, 151, 153–5
    ideal, 151–3
    moveable, 151, 155–9

Platinum resistance thermometers, 34
Platinum-rhodium thermocouples, 34
Pneumatic devices, 114
Polarographic oxygen electrodes, 125
Potential, electrical, 171
Potential temperature, 38
Potentiometers, 30, 176
Power
   electrical, 171
   shipboard, 4
Power supply, buoy, 154
Probable error, 11
Propellers, 70
Propagation loss, 118
Precision, 14
Pressure, 4, 26
   aneroid element sensor, 28
   Bourdon tube sensor, 82
   capacitive sensor, 29
   carbon sensor, 29
   quartz sensor, 29
   range, 31
   resistive sensor, 29
   sensors, 27–20
   sensor problems, 30–1
   thermometer sensor, 27
   tunnel-diode sensor, 29
   units, 27
   Vibratron sensor, 29
   wave sensor, 137
Pressure gradient force, 62
Pressure plate current meter, 73
Punched-paper tape, 72
Pyrometers, 40, 42–3
PZT, 115

Quartz, 29, 31
Quinhydrone pH electrodes, 127

Radio data links, 166
Random error, 13
Random variations, 10
R-C circuits, 15
Readout, definition, 6
Reed switches, 71
Reflectance, 91–2
Refraction, sound, 108
Reliability, 167

Repeatability, definition, 14
Resistance, electrical, 171
   measurement, 176
Resolution, 22
Resonant circuits, 175
Reverberation, 107
Reynolds stress, 62

Salinity
   conductivity sensors, 50–3
   definition, 47–8
   density relationship, 48
   distribution, 49
   instrument system, 6
   laboratory determination, 49
   microwave measuring method, 54–6
   refractometer measuring method, 50
   sound-speed measuring method, 54
Salinity-brightness temperature, 56
Salinity-conductivity relationship, 48
Salinometers, 49–53
Satellite
   current measurements, 77
   data telemetry, 167
   depth measurements, 25
   instrument platforms, 158–9
   salinity measurements, 54–6
   temperature measurements, 42
   wave measurements, 141
Savonius rotors, 70
Scatter, 10
Scattering
   light, 83, 85
   sound, 106
Scattering coefficient, light, 84
Scuba, 158
Sea, 133, 135
Sea-floor spreading, 5
Sea-surface noise, 112
Secchi disc, 98
Seismic sea waves, 136
Sensitivity, definition, 22
Sensors
   current, 68–77
   definition, 5
   gravity, 145–6
   light, 88, 94, 95

magnetic-field, 148
oxygen, 125–6
pH, 127–8
pressure, 27–31
salinity, 49–56
sound, 114–7
temperature, 40–3
tide, 141
wave, 137–41
Sensor response, 15
Series circuit
  resistance, 173
  inductance, 173
  capacitance, 173
Shadow zones, 110
Shallow-water waves, 132–3
Ships
  catamarans, 156
  drift of, 66, 69
  submersibles, 156–8
  surface, 155–6
Shipping noise, 112
Signal conditioners, 6, 30, 72
Sing-around circuit, 119
Singing wire, 76
Single-conductor cables, 165
Sonic current meters, 76
Sonic scattering layer, 107
Sound
  absorption, 105–6
  holography, 111–2
  noise, 112–3, 118
  refraction, 108–11
  propagation loss, 118
  scattering, 106–7
  sources, 114
  speed measurement, 119–20
  spreading, 107–8
  time-dependent losses, 111
  transducers, 114–7
  uses, 105
Sound-propagation speed, 108
Sound velocimeters, 119–20
Sources, sound, 114
SPAR, 153
Spar buoys, 153
Sparkers, 114
Spectroradiometers, 94
Spectra

real-time, 141
water-wave, 133
Spherical spreading, 108
Splashnik, 141
Spread, 10
Spreading, 107–8
Spring bellows, 28
SSL, 107
Stable buoys, 153
Staffs, wave, 139
Standard atmosphere, 27
Standard deviation, 10
Standard error, 11
Static error band, 22
Statistical parameters, 9–12
Stefan-Boltzmann law, 42
Stereo-photography, wave, 141
STD, 52, 53
Strain gages, 30
Student's t-distribution, 12
Submersibles, 156–8
Surface-following buoys, 153
Survey ships, 155–6
Swallow floats, 68
Swell, 133, 135
Synoptic data, 151, 159
Systematic error, 13

t-distribution, 12
Taut-line moorings, 154
Temperature
  accuracy, 39
  definition, 33
  estuarine variation, 39
  oceanic, 35–8, 39
  potential, 38
  primary points, 34
  scales, 34–5
  sensors, 40–3
Thallium oxygen electrode, 126
Theoretical transfer function, 22
Thermistors, 41
Thermistor beads, 137
Thermistor chains, 137
Thermoclines, 19, 109
  diurnal, 37
  main, 36
  seasonal, 37
Thermometers, 40

capillary, 40
infrared, 42
liquid in glass, 40
protected, 27
quartz, 42
resistance, 41
reversing, 27
thermistor, 41
unprotected, 27
Threshold, 23
Thumpers, 114
Tidal currents, 63
Tide gages, 141
Tide waves, 136
Time constant
 current sensors, 66
 depth sensors, 31
 mathematical development, 14–22
 oxygen electrode, 126
Toroidal conductivity sensors, 51–3
Tracers, 68
Trade-offs, 7
Transducers (*see also* Sensors)
 acoustic, 114
 calibration, 116
 definition, 5
 requirements, 117
 tests, 117
Translators, definition, 6
Transmission anomalies, 118
Transmissometers, 95–8
Trieste, 26
Tsunami, 136
Tunnel diodes, 29
Turbulence, 5, 61, 111
 measurement, 76–7
Two-point moorings, 154

Uncertainty, percent, definition, 11
Undercurrents, 63
Units, current, 77

Van Dusen's formula, 34
Variable-frequency oscillators, 30
Variance, 10
Vector averaging, 78
Vertical motion, 64
Vibratron, as instrument problem, 4
Vibration, 29
Visibility, 4, 94
Volt, definition, 171
Voltage, measurement, 176
Vortex shedding, 76, 162
Vortex street, von Karman, 76

Water
 density, 4
 light-loss characteristic, 85
 sound-speed characteristics, 108–9
Watt, definition, 172
Wavelength, sound, 106
 surface waves, 133
Waves, surface
 deep, 131–3
 height, 136
 measurements, 135–6
 periods, 133
 pressure, 132
 sensors, 137–42
 shallow, 132
 spectra, 133–5
 speed, 131–2, 133
 staffs, 139
 wavelength, 133
Wheatstone bridge, 176
Winkler titration, 125
Wire angle, 25
Worden gravimeters, 146
Wratten filters, 91

XBT, 41–2
XSTD, 54, 55, 57

Composed in ten-point Times Roman with
three points of leading by Monotype Composition Company, Baltimore, Maryland.

Printed offset on sixty-pound Finch Tradebook Offset
and bound in Holliston Roxite C, Vellum
Finish, by The Maple Press, York,
Pennsylvania.

Line drawings by William J. Clipson